Carbon Inequality

With a specific focus on the United States and the United Kingdom, *Carbon Inequality* studies the role of the richest people in contributing to climate change via their luxury consumption and their investments. In an innovative contribution, it attempts to quantify personal responsibility for shareholdings in large fossil fuel companies.

This book explores the implications of the richest people's historic responsibility for global warming, the impacts of which affect them less than most others in global society. Kenner analyses how the richest people running large oil and gas companies have successfully used their political influence to lobby the US and UK government. This assessment of their growing political power is particularly pertinent at a time of increasing inequality and growing public awareness of the impact of climate change. The book also highlights the crucial role of the richest in blocking the low-carbon transition in the US and the UK, exploring how this could be countered to ensure fossil fuels are fully replaced by renewable energy.

This book will be of great relevance to scholars and policy makers with an interest in inequality, climate change and sustainability transitions.

Dario Kenner is a visiting fellow at the Global Sustainability Institute based at Anglia Ruskin University, UK. This book builds on his 2015 working paper *Inequality of Overconsumption: the Ecological Footprint of the Richest*, published by Anglia Ruskin University.

Routledge Focus on Environment and Sustainability

Indigenous Sacred Natural Sites and Spiritual Governance
The Legal Case for Juristic Personhood
John Studley

Environmental Communication Among Minority Populations
Edited by Bruno Takahashi and Sonny Rosenthal

Solar Energy, Mini-grids and Sustainable Electricity Access
Practical Experiences, Lessons and Solutions from Senegal
Kirsten Ulsrud, Charles Muchunku, Debajit Palit and Gathu Kirubi

Climate Change, Politics and the Press in Ireland
David Robbins

Productivity and Innovation in SMEs
Creating Competitive Advantage in Singapore and Southeast Asia
Azad Bali, Peter McKiernan, Christopher Vas and Peter Waring

Climate Adaptation Finance and Investment in California
Jesse M. Keenan

Negotiating the Environment: Civil Society, Globalisation and the UN
Lauren E. Eastwood

Carbon Inequality
The Role of the Richest in Climate Change
Dario Kenner

For more information about this series, please visit: https://www.routledge.com/Routledge-Focus-on-Environment-and-Sustainability/book-series/RFES

Carbon Inequality

The Role of the Richest in Climate Change

Dario Kenner

First published 2019
by Routledge
2 Park Square, Milton Park, Abingdon, Oxon OX14 4RN

and by Routledge
605 Third Avenue, New York, NY 10017

First issued in paperback 2020

Routledge is an imprint of the Taylor & Francis Group, an informa business

British Library Cataloguing-in-Publication Data
A catalogue record for this book is available from the British Library

Library of Congress Cataloging-in-Publication Data
Names: Kenner, Dario, author.
Title: Carbon inequality: the role of the richest in climate change / Dario Kenner.
Description: Abingdon, Oxon; New York, NY: Routledge, 2020. |
Series: Routledge focus on environment and sustainability | Includes index.
Identifiers: LCCN 2019014361 (print) | LCCN 2019017517 (ebook) |
ISBN 9781351171328 (Master) | ISBN 9780815399223 (hardback) |
ISBN 9781351171328 (ebook)
Subjects: LCSH: Wealth—Environmental aspects. |
Consumption (Economics)—Environmental aspects. |
Affluent consumers. | Climatic changes—Social aspects.
Classification: LCC HC79.W4 (ebook) |
LCC HC79.W4 K457 2020 (print) | DDC 363.738/74—dc23
LC record available at https://lccn.loc.gov/2019014361

ISBN 13: 978-0-367-72766-6 (pbk)
ISBN 13: 978-0-8153-9922-3 (hbk)

Typeset in Times New Roman
by codeMantra

Contents

Figures

Tables

Acknowledgements

This book is dedicated to Georgina and our two daughters. Her unwavering support and insightful feedback made this book possible.

My father José accompanied me throughout the research and writing. His constant difficult questions and valuable suggestions pushed me to go further and develop my ideas.

My mother Charmian played a crucial role in the final stages with her logic and attention to detail.

I would also like to thank James Boyce, Tina Fawcett, Simon Pirani and Max Lawson for important comments on drafts. I would also like to thank Richard Heede for sharing ideas on the database.

London, March 2019

Introduction

Climate change and the role of the richest

High and rising inequality is concentrating wealth and political power in the richest people around the world. This trend is clearly present in the United States and the United Kingdom, two countries with large historical responsibility for global warming. In the age of climate change and the sixth mass extinction, there is a situation of carbon inequality where the richest people in the US and the UK have an *unequal ability to pollute*. This is via their high-carbon luxury consumption and their *investment emissions*, when they hold shares in companies that produce greenhouse gas emissions.

Wealthy shareholders who are also decision-makers at large multinational oil, gas and coal companies form part of the *polluter elite*, a group I begin to identify in a database that accompanies this book. Based on the size of their investment emissions I conclude that these decision-makers, such as the executive team and directors, hold greater historical responsibility for climate change. It is time to recognise their unique role in the Anthropocene. There is also an environmental injustice as they suffer fewer of the consequences of pollution from their consumption and investments.

The polluter elite seek to obtain political influence over the state to ensure the profitability of their shareholdings which are the basis of their personal net worth. This is as decision-makers who approve lobbying by their companies, and in some cases in a personal capacity by using their own wealth to donate to political parties. The political influence of the polluter elite has been decisive in slowing down the transition away from fossil fuels over the past few decades. They have exerted their power over the state which depends on fossil fuels to spur economic growth for its legitimacy. They have formed alliances with policy makers to undermine policies that would reduce greenhouse gas emissions. The polluter elite has thus ensured that despite the governments of the US and the UK acknowledging the danger of global warming since the early 1990s, both economies remain heavily reliant on fossil fuels.

Increasing scientific evidence of global warming requires the transition away from fossil fuels to be deliberately accelerated. This will only happen if the polluter elite are undermined through policies such as the removal of fossil fuel subsidies, a carbon tax and the phasing out of fossil fuels. Overwhelming public pressure could counter the political influence of the polluter elite and force policy makers into action to ensure fossil fuels are replaced by renewable energy.

1.1 Context: global warming and species extinction

The world faces significant environmental challenges epitomised by alarming rates of biodiversity loss and the heightened prospect of irreversible climate change. Global populations of fish, birds, mammals, amphibians, and reptiles declined by 60% between 1970 and 2014. The top threats to species are directly linked to human activities such as habitat loss and overexploitation of wildlife. Invasive species, pollution and climate change are also factors.[1] The latest assessments by the Intergovernmental Science-Policy Platform on Biodiversity and Ecosystem Services (IPBES) show that biodiversity is declining rapidly in every region in the world. The report highlights that biodiversity is under threat from the overexploitation and unsustainable use of natural resources and related air, land and water pollution.[2] Land degradation, intimately linked with biodiversity loss and a driver of climate change, is having a negative impact on the well-being of over 3 billion people.

There have been five previous mass extinctions in the last 540 million years. A range of studies show there are signs that we are on the verge of the sixth mass extinction as current extinction rates are higher than the long-term rate, known as the background extinction rate.[3]

With regard to climate change, greenhouse gases that occur naturally in the atmosphere include carbon dioxide (CO2), methane (CH4) and nitrous oxide (N2O). However, these key greenhouse gases have increased significantly due to human activity. They are making the planet warmer by absorbing infrared radiation and trapping heat. This means that we face the prospect of runaway climate change which will see an increase in extreme weather events including hurricanes, flooding, forest fires and drought, as well as the melting of glaciers which will affect access to water.[4]

Since 1990 the Inter-Governmental Panel on Climate Change (IPCC) has published reports recording higher levels of greenhouse gases and detailing their impacts. In 2018 the organisation released a special report on the impacts of global warming of 1.5°C above pre-industrial levels.[5] Its main conclusions were that climate change is a result of human activity and that limiting global warming requires deep emissions cuts by 2030. It also recommends a rapid shift towards renewable energies. Key headlines from the report include (Table 0.1):

Due to space limitations the focus in this book is principally on climate change. That said it is important to place the debates and ideas presented in this book in the context of the broader environmental challenges such as large-scale levels of biodiversity loss and acceleration in

Table 0.1 Selected key messages from the 2018 IPCC report

Headlines	*Implications*
"Human activities are estimated to have caused approximately 1.0°C of global warming above pre-industrial levels, with a likely range of 0.8°C to 1.2°C. Global warming is likely to reach 1.5°C between 2030 and 2052 if it continues to increase at the current rate"	Emissions need to fall rapidly by 2030 to prevent an increase of 1.5°C above pre-industrial levels.
"Limiting global warming requires limiting the total cumulative global anthropogenic emissions of CO_2 since the preindustrial period, i.e. staying within a total carbon budget" "Estimated anthropogenic global warming is currently increasing at 0.2°C (likely between 0.1°C and 0.3°C) per decade due to past and ongoing emissions"	Emissions are cumulative meaning that continued emissions, even one more tonne of carbon dioxide, at this point are dangerous because together they trigger temperature rises.
"Future climate-related risks depend on the rate, peak and duration of warming" "Some impacts may be long-lasting or irreversible, such as the loss of some ecosystems"	There are tipping points e.g. melting permafrost.
"On land, impacts on biodiversity and ecosystems, including species loss and extinction, are projected to be lower at 1.5°C of global warming compared to 2°C"	Climate change is one driver of the sixth mass extinction.
"Pathways limiting global warming to 1.5°C with no or limited overshoot would require rapid and far-reaching transitions in energy, land, urban and infrastructure (including transport and buildings), and industrial systems. These systems transitions are unprecedented in terms of scale, but not necessarily in terms of speed, and imply deep emissions reductions in all sectors, a wide portfolio of mitigation options and a significant upscaling of investments in those options"	The more that business as usual continues the more drastic the emissions cuts that are needed.[6]

Source: IPCC, 2018.

the rate of species extinction. Many of the drivers are the same (e.g. extraction of fossil fuels) with overlapping impacts (e.g. global warming increases the risks that more species will become extinct).

The issue of global warming rose up the international political agenda in the late 1980s. Following the formation of the IPCC in 1988 one of the key outcomes of the 1992 Earth Summit was the creation of the United Nations Framework Convention on Climate Change (UNFCCC). The objective of the UNFCCC was to stabilise greenhouse gas concentrations in the atmosphere to prevent dangerous human interference with the climate system (UNFCCC website). Twenty-one Conferences of the Parties (COPs) to the convention were held since 1994 leading to the Paris Agreement at the COP held in Paris in December 2015. To date 184 countries have ratified the agreement. Under the Paris Agreement governments committed to keep global temperatures "well below" 2.0C (3.6F) above pre-industrial times and "pursue efforts to limit the temperature increase to 1.5C".[7] The negotiations continue to establish rules and review mechanisms to monitor progress.

Whilst the Paris Agreement has pushed forward the climate change agenda by deepening the scientific consensus on the need to reduce emissions it is unlikely to meet its own targets. Based on current trends there is an emissions gap that will see the global carbon budget completely used up by 2030 meaning it will be difficult to limit global warming to below 2°C.

1.2 Who is responsible for climate change?

The IPCC reports make it clear that climate change is caused by human activity. The logical next question is which humans are responsible? To date debates have mainly revolved around the differences in emissions between countries. This book contributes to these existing debates by looking at what is happening within countries. Do some citizens have larger emissions associated with their consumption and investments than others? At a time of high and increasing inequality in the United States and the United Kingdom this is a valid question. Average figures such as income and emissions per capita often hide extreme differences in how both are distributed across the population.

There are several reasons I have chosen to focus on the US and UK:

- **They both have large historical responsibility in terms of their greenhouse gas emissions**: The Industrial Revolution began in the UK and it used coal on a large scale. It is estimated that the UK was responsible for 80% of global emissions by 1825 and 62% of global emissions by 1850.[8] Overall the US has had an enormous environmental impact as the largest economy and largest consumer of natural resources in the world. This makes it crucial to understand the role of the richest in shaping environmental outcomes in the US and globally. In the US many of the largest fortunes were based on oil and automobiles from the 1890s onwards.

- **They are two key countries in terms of the global concentration of wealth**: Research on the country of residence of High Net Worth Individuals (HNWIs) shows the US and the UK are key locations. HNWIs are people with a "minimum of US$1 million in investable wealth, excluding primary residence, collectibles, consumables, and consumer durables".[9] According to the World Wealth Report in 2017 the US had the most HNWI residents (5.2 million), Japan (3.1 million), Germany (1.3 million), China (1.2 million), France (629,000) and in sixth place was the United Kingdom (575,000). While the Capgemini study, and other rich lists, are by no means completely accurate, they are a starting point in tracking the location of the richest people.[10] In the case of Ultra High Net Worth Individuals, who have at least $30 million in investable wealth, they mainly live in London and New York.[11] London in particular is a favourite destination for the global super-rich.
- **The richest are capturing more and more income and wealth in both countries**: This has led some observers to argue there is a trend of plutonomy in countries such as the US and UK.[12] Plutonomy refers to a situation where the richest are the group that holds the highest percentage of wealth, receives the highest income and consumes more than other groups (Figures 0.1 and 0.2).

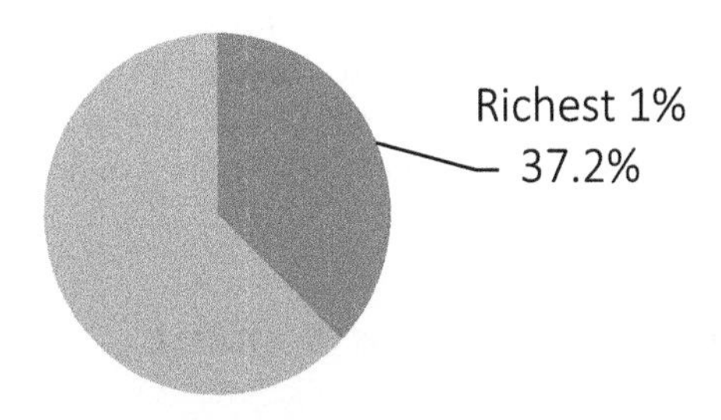

Figure 0.1 Percentage of wealth held by the richest 1% in the US in 2014.
Source: World Inequality Database, 2018.[13]

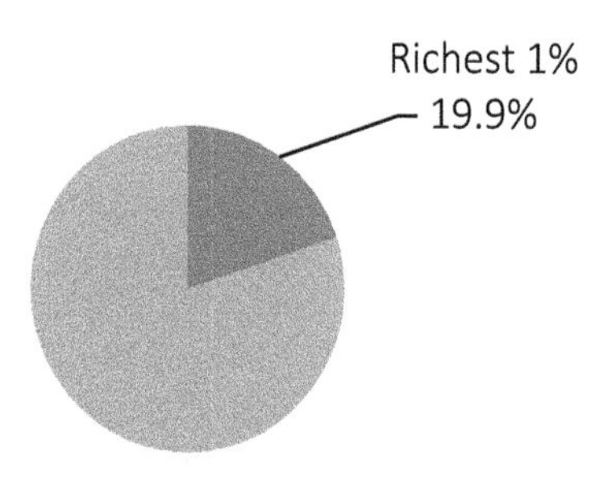

Figure 0.2 Percentage of wealth held by the richest 1% in the UK in 2012.
Source: World Inequality Database, 2018.[14]

It is true that in a globalised economy, individuals are highly mobile, and this is particularly the case for the richest. Focusing on the US and UK is a starting point in deepening understanding of the role of the richest to advance discussion of this issue. It is not my intention to claim this as the only way to approach this question or to close off other avenues of research which could combine the national with the international – which would reflect the globalised nature of the consumption and investments of the richest people. Clearly, there are particular dimensions to the unequal contribution to climate change in every country around the world.[15] My hope is this book will spur more country specific research which will build a global picture of the role of the richest people in contributing to global warming.

The reader may ask why I am focusing on a rich minority when climate change is a systemic issue that needs a society-wide response (which would ideally be coordinated at the multilateral level building on the Paris Agreement). Clearly it is not only the richest people who pollute and therefore the pollution of the rest of the population should not be ignored. However, I argue with high levels of inequality it is time to recognise and increasingly scrutinise what I call the *unequal ability to pollute* in the age of climate change. Indeed, one of the objectives of this book is to highlight the ability of the richest people to have influence at the political and economic level over emissions from production and consumption.

1.3 Who do I mean when I discuss the richest people in the US and UK?

There are a number of ways to identify the richest individuals in a country. When I began this research, it became clear that depending on the rich individual they had different dimensions to their role in climate change. Together they are the basis of the *unequal ability to pollute*. The key areas are:

1 the consumption carbon footprint,
2 emissions associated with investments,
3 political influence.

Whilst it is reasonable to state that all rich individuals in the US and the UK have a carbon footprint from their consumption, it is not true that all rich individuals invest in oil companies, let alone that they all personally donate to political parties. For consumption emissions I focus on the richest 1% *by income* in the US and the UK. For investment emissions and political influence, I look at a group I call the *polluter elite*. These are decision-makers such as the CEO and directors at oil and gas companies. Some, not all, members of this polluter elite are captured in the polluter elite database (discussed further below).

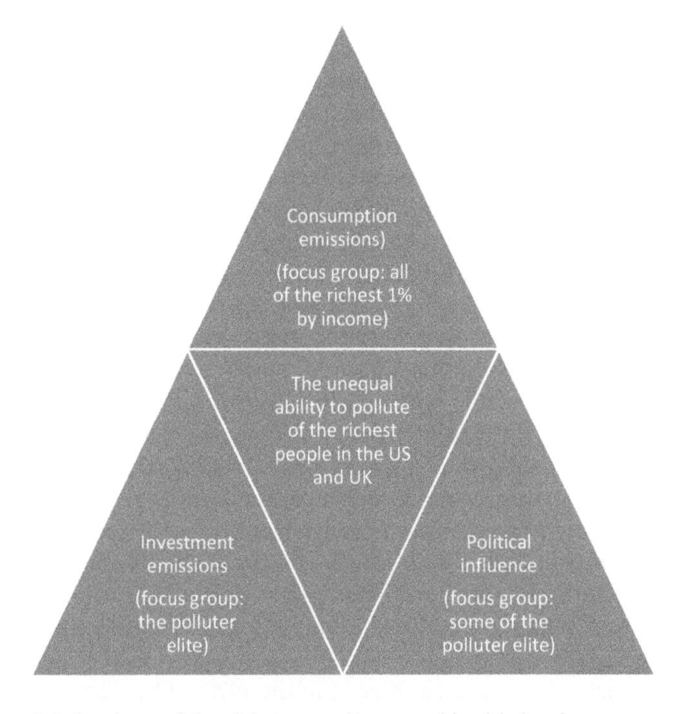

Figure 0.3 Sections of the richest people covered in this book.
Source: Author.

This is partly due to confidentiality surrounding the investments of the richest and a lack of resources as it would be an ambitious project to document all of the polluter elite. The objective of the polluter elite database is to illustrate trends rather than be a comprehensive list of all members of the polluter elite. The full database is available at: https://whygreeneconomy.org/

Wealth is the foundation of the unequal ability to pollute. As the wealth of the richest increases so does their ability to pollute. The concentration of wealth enables them to increase their consumption, investments in polluting companies, and donations to political parties (Figure 0.3).

1.4 Time to study the rich – An uncommon exercise that needs to become mainstream

In the 1950s Mills published his pioneering and provocative book on the power elite in the US. He demonstrated the close interlocking between the groups running political, economic and military institutions. He claimed unprecedented power was being concentrated in their hands in a process of centralisation of decision-making. He argued their power was based not just on having amassed more wealth or prestige, but also rested on their ability to "realise their will, even if others resist it".[16]

To date the richest have often been neglected in academic study. In 1972 the anthropologist Nader made a call to "study up" as well as "down" to understand how power is exercised (her focus was the US). Nader noted there was very little research on the upper classes,[17] partly because the powerful can be difficult to access, and therefore argued for a shift from asking why there were poor people to ask why others were more wealthy.[18] She declared that: "the study of man is confronted with an unprecedented situation: never before have so few, by their actions and inactions, had the power of life and death over so many members of the species". It is in this spirit that this book examines the role of the richest people in climate change.[19]

Nader's call is still just as relevant today. In the years leading up to the 2008 financial crisis the ever-higher earnings of the global elite went largely unnoticed in academic research, which was surprising given widening inequality and their increasingly conspicuous consumption based on luxury mansions, yachts, exclusive country clubs and private islands.[20]

Savage and Williams[21] argue it is fundamental to re-focus on elites, something which became unfashionable across the social sciences after a brief period of research interest in the 1970s. There are now several new areas being explored in elite studies. They include the use of new databases, new methods such as social network analysis, and new topics such as gender and the role of elites in the financial sector. This has led Savage and Nichols to conclude that the sociological analysis of elites has been rejuvenated, even if the revival has been uneven.[22] Such studies complement and go beyond the increasing focus on inequality by economists who concentrate on issues of income and wealth.

At a time of deepening inequality, and as evidence builds of global warming taking place, the unequal ability to pollute of the richest people, and in particular their efforts to influence the state to favour fossil fuels, deserves greater academic scrutiny. It is much more common to study the impact of climate change policy on the poor than on the richest.[23] It is time to look in more detail at who is responsible for greenhouse gas emissions and who is blocking the shift to a low-carbon economy.

1.5 Overview of the book

Chapter 1 looks at the luxury consumption of the richest people in the US and the UK. Several studies confirm they have higher carbon footprints based on activities such as flying. Their desire to be hypermobile is linked to status competition. Their high-carbon lifestyle is one component of what I call the *unequal ability to pollute*. The other component is their *investment emissions* which are dealt with in Chapter 2. This chapter presents the findings from a database which compiles information on several hundred wealthy individuals (mainly the executive team and directors) with shareholdings in the oil and gas companies with the largest historical emissions. They are members of a group I call the *polluter elite*. As decision-makers at these companies these individuals hold responsibility for the direct emissions from their companies' operations, and therefore more historical responsibility for climate change. These findings contribute to existing debates which tend to compare emissions between countries rather than look at the unequal distribution of emissions within countries. The large investment emissions of the polluter elite mean they play a unique role in the Anthropocene.

Chapter 3 raises pertinent moral questions. Why is it the polluter elite are more able to avoid the consequences of their pollution? Is it morally questionable to profit from pollution when it can have a fatal impact via air pollution and extreme weather events? The chapter questions the status of the polluter elite as wealth creators and proposes the pollution of the richest via their investments is recognised when discussing their vast fortunes. I propose new rich lists that transparently show when a rich individual has made their money from polluting activity. This would recognise the unequal ability to pollute and the specific role of the polluter elite in contributing to global warming.

Chapter 4 documents the political activity of the polluter elite to understand how they influence policy makers. It argues the polluter elite have successfully exploited the state's dependency on fossil fuels to increase economic growth. The evidence of their influence is clear. Since the 1990s successive governments have continued support for polluting companies through legislation and subsidies. This has happened despite increasing scientific evidence of global warming and the urgency to reduce emissions. At a time of rising inequality, the elite's political power is growing as economic resources are further concentrated in their hands.

Given the dominant political power of the oil & gas polluter elite the low-carbon transition will only happen on the large scale and at the rapid speed that is needed if they are weakened. Chapter 5 discusses how this could happen using lessons learnt from the reduction in coal production and use in both countries. It assesses the political and economic factors that could lead to the rapid destabilisation of the oil & gas companies. There is a particular focus on the key role of public pressure in influencing whether policy makers will act to reduce the dominance of oil and gas in the economies of the US and UK.

Notes

1 World Wildlife Fund (WWF), 2018. Living Planet Index, [online]. *WWF*. Retrieved from: https://c402277.ssl.cf1.rackcdn.com/publications/1187/files/original/LPR2018_Full_Report_Spreads.pdf [Accessed 20 December 2018].

2 Intergovernmental Science-Policy Platform on Biodiversity and Ecosystem Services (IPBES), 2018. Biodiversity and Nature's Contributions Continue Dangerous Decline, Scientists Warn. IPBES Media Release. Retrieved from: www.ipbes.net/news/media-release-biodiversity-nature%E2%80%99s-contributions-continue-%C2%A0dangerous-decline-scientists-warn [Accessed 20 May 2018].

3 Gerardo Ceballos, G., Ehrlich P. R. and Dirzo, R., 2017. Biological Annihilation Via the Ongoing Sixth Mass Extinction Signalled by Vertebrate Population Losses and Declines. *PNAS*, [online], 114 (30). Retrieved from: doi:10.1073/pnas.1704949114 [Accessed 20 May 2018].

4 Inter-Governmental Panel on Climate Change (IPCC), 2013. Climate Change 2013: The Physical Science Basis. *Fifth Assessment Report*. Retrieved from: www.ipcc.ch/site/assets/uploads/2018/02/WG1AR5_SPM_FINAL.pdf [Accessed 20 May 2018].

5 Inter-Governmental Panel on Climate Change (IPCC), 2018. Global Warming of 1.5°C. Retrieved from: https://report.ipcc.ch/sr15/pdf/sr15_spm_final.pdf [Accessed 20 November 2018].

6 Larkin, A., 2018. What If Negative Emission Technologies Fail at Scale? Implications of the Paris Agreement for Big Emitting Nations. *Climate Policy*, [online], 18 (6). Retrieved from: doi:10.1080/14693062.2017.1346498 [Accessed 20 May 2018].

7 United Nations Framework Convention on Climate Change (UNFCCC). 2015. Paris Agreement. *UN*. Retrieved from: http://unfccc.int/files/essential_background/convention/application/pdf/english_paris_agreement.pdf [Accessed 20 July 2018].
8 Bonneuil. C. and Fressoz, J-B., 2017. *The Shock of the Anthropocene: The Earth, History and Us*. London: Verso.
9 Capgemini, 2018. HNWI Market Expands. Capgemini. Retrieved from: https://worldwealthreport.com/hnwi-market-expands/ [Accessed 20 November 2018].
10 Piketty, T., 2014. *Capital in the Twenty-First Century*. Cambridge, MA: Harvard University Press.
11 Wealth-X., 2017. HNWI Market Expands. Wealth-X. Retrieved from: https://worldwealthreport.com/hnwi-market-expands/ [Accessed 20 November 2018].
12 Hay, I., 2016. On Plutonomy: Economy, Power and the Wealthy Few in the Second Gilded Age. *In*: Hay, I. and Beaverstock, J., eds. *Handbook on Wealth and the Super-Rich*. Cheltenham: Elgar, 77; Makdissi, P. and Yazback, M., 2015. On the Measurement of Plutonomy. *Social Choice and Welfare*, [online], 44(4). Retrieved from: https://link.springer.com/article/10.1007/s00355-014-0857-0 [Accessed 12 May 2018].
13 World Inequality Database, 2018. Wealth Inequality, USA, 1970–2014. *World Inequality Database*. Retrieved from: https://wid.world/country/usa/ [Accessed 20 November 2018].
14 World Inequality Database, 2018. Top 1% Net Personal Wealth Share, United Kingdom, 1895–2014. *World Inequality Database*. Retrieved from: https://wid.world/country/united-kingdom/ [Accessed 20 November 2018].
15 China has the fastest growing HNWI population (Capgemini, 2017). However, it is very difficult to access data on income and wealth distribution, as well as environmental impact.
16 Mills, C. W., 1959. *The Power Elite*. New York: Oxford University Press.
17 With the notable exceptions of Lundberg's work on America's richest families (1937) and Mills on the Power Elite (1956).
18 Nader, L., 1972. Up the Anthropologist: Perspectives Gained From Studying Up. *In*: Hymes, D., ed. *Reinventing Athropology*. New York: Pantheon Books, 284, 289, 302.
19 Kempf, H., 2008. *How The Rich Are Destroying the Earth*. Cambridge: Green Books.
20 Caletrío, J., 2012. Global Elites, Privilege and Mobilities in Post-organized Capitalism. *Theory, Culture and Society*, [online], 29 (2). Retrieved from: doi:10.1177%2F0263276412438423 [Accessed 4 May 2018].
21 Savage, M. and Williams, K., eds. 2008. *Remembering Elites*. Oxford: Wiley-Blackwell.
22 Savage, M. and Nichols, G., 2017. Theorising Elites in Unequal Times: Class, Constellation and Accumulation. *In*: Korsnes, R. et al., eds. *New Directions in Elite Studies*. London: Routledge, 7.27–7.42.
23 Otto, L. et al., 2019. Shift the Focus from the Super-poor to the Super-rich. *Nature Climate Change*, [online], 9 (82–84). Retrieved from: www.nature.com/articles/s41558-019-0402-3 [Accessed 17 February 2019].

1 The carbon footprint of luxury consumption

When I began my research on the environmental impact of the richest people (see Kenner, 2015), their luxury lifestyle was an intuitive place to start. Images of wealthy people flying around the world in their private jets to mansions are probably at the forefront of the popular imagination when thinking about the role of the richest in climate change.

This chapter will argue that, per person, rich individuals consume in more high-carbon ways compared to other citizens. As many of the existing studies on this area calculate the carbon footprint of the richest 1% by income, I have concentrated on this group of the richest people in the US and the UK. The focus here is personal direct and indirect emissions, for example, from driving, flying or owning several luxury cars. This does not address emissions associated with business travel and hospitality, which, while a crucial area to deal with, are beyond the scope of this book.

1.1 Luxury lifestyles

The *unequal ability to pollute* is not a new phenomenon. Wealth has always determined access. In the US of the 1930s cars were luxury goods and were very expensive until they became more widely owned from the 1940s. There was a similar trend in Europe from the 1950s to the 1960s. Similarly, flying started as an elite mode of transportation. After the Second World War the airlines opened up flying to more passengers but it was still the richest who flew the most, epitomised by the use of Concorde from the mid-1970s. In the period prior to mass consumption of these high-carbon forms of transport it was the rich minority in the US and the UK who held greater historical responsibility for consumption-based emissions from cars and flying.

Moving to the present day there are many examples of how these lifestyles are high-carbon. Several websites give an insight into the consumption patterns of the richest, such as the Forbes *Cost of Living Extremely Well Index*,[1] and *How to Spend It*.[2] What these sources reveal is that the richest live in large luxurious residences (with some of the very rich owning their own islands) and own luxury cars, yachts, private jets and submarines. Social media provides a graphic insight into the extravagant lifestyles of wealthy young people, including their use of private jets and luxury cars. Their social calendar includes exclusive events such as the annual Davos World Economic Forum and the Dubai International Boat Show. These emblematic examples confirm academic research which finds a correlation between the level of income and the size of carbon footprint.[3]

Moving from these symbolic examples, is it possible to quantify the high-carbon lifestyle of the richest people in the US and the UK? Several academic studies estimate the carbon footprint of the richest income group in each of these countries. In November 2015 Chancel and Piketty published a paper exploring carbon inequality in countries around the world.[4] They use their data to obtain global estimates of carbon inequality. They argue it is important to look at what is happening with high emitters within countries as well as between countries. Chancel and Piketty focus on consumption-based emissions. This attributes emissions to the final consumers rather than the companies producing these goods and services. They have a particular emphasis on indirect emissions (e.g. from goods such as food and services such as hotels). Their methodology is based on national level data, including per capita CO2e emissions (carbon dioxide equivalent), consumption-based CO2e emissions, and income inequality covering the period 1998–2013. They stress that their data should be treated with care and not as definitive. Therefore, the results presented below for the US and the UK should be seen as estimates of what could be happening rather than the actual figures of each income group.

Chancel and Piketty estimate that the top 10% of emitters in all countries account for around 45% of global emissions. Their data shows that in the US and other high-emitting countries such as the UK, Australia, Canada, China, France, Germany, Mexico, India, Indonesia, Russia, Saudi Arabia and Japan the richest 1% by income (in each of these countries) have much higher average CO2e emissions per person. Whilst acknowledging the diversity of national contexts it is clear that, the issue of carbon inequality is relevant to a range of countries. For example, in Nigeria it is estimated that rich individuals who made their money from oil spent $6.5 billion on private jets between 2007 and 2012.[5]

The country where the people making up the richest 1% are estimated to have the highest per capita carbon footprint is the US. Chancel and Piketty calculate that in 2013 the average emissions per person of the richest 1% (3.2 million people) were around 318 metric tons of CO2e. In comparison the average emissions per person of the poorest 10% (around 31 million people) were around 3.6 metric tons of CO2e. For the top 1%, Chancel and Piketty estimate 55 metric tons per person from direct emissions such as transport and household energy. The rest comes from indirect emissions of which investments are likely to be a large part. In Chapter 2 of this book, I present evidence on emissions associated with investments. This indicates that Chancel and Piketty are correct in their intuition that the largest portion of personal emissions of the richest people comes from the shares they hold.

Chancel and Piketty's data shows that the unequal distribution of greenhouse gas emissions has likely deepened over the last 15 years. The only income group which saw their carbon footprint increase between 1998 and 2013 was the richest 1% whose average annual emissions per person rose from 289 to 318 metric tons of CO2e. Meanwhile, the remaining 99% saw their average annual emissions per person fall slightly over the period 1998 to 2013. This applies to all of the 99%, although clearly there are significant differences in the size of footprint (see Table 1.1). It is interesting to note that carbon inequality is estimated to have got worse in the US at the same time as income and wealth inequality dramatically increased, most notably in the hands

Table 1.1 Per capita consumption-based emissions in the US in 2013

United States	*Income group (%)*	*Consumption based CO2e emissions per capita 2013*
Richest	1	318.3
	9	58.5
	10	32.6
	10	25.6
	10	21.1
	10	17.6
	10	14.8
	10	12.2
	10	9.7
	10	7.1
Poorest	10	3.6

Source: Chancel and Piketty, 2015.

of the richest 1%. Therefore, it would appear there is a link between rising inequality and carbon inequality.

In the UK, Chancel and Piketty estimate that in 2013 the richest 1% of people (around 64,000 people) were responsible for 147 metric tons of greenhouse gas emissions per person. In comparison the poorest 10% (around 6.4 million people) were responsible for around 4 metric tons of greenhouse gas emissions per person in 2013 (Table 1.2).

In December 2015 Oxfam published a report on Extreme Carbon Inequality using a similar methodology to that of Chancel and Piketty.[6] In it they estimate that someone who is in the richest 1% by income in the world could have an average footprint as much as 175 times that of someone in the poorest 10%. They calculate that the richest 10% are responsible for around 50% of global emissions compared to just 10% for the poorest 50%. In a more recent study Hubacek et al call for more focus on the link between economic inequality and high-carbon lifestyles.[7] They estimate that globally the richest 10% by income were responsible for an estimated 36% of emissions as a result of goods and services they consumed. This factored in the production process all the way along global supply chains. In a more recent study a typical high wealth household was estimated to have an annual carbon footprint of 129.3 metric tons of CO2e, with the main source of emissions being air travel.[8]

Table 1.2 Per capita consumption-based emissions in the UK in 2013

United Kingdom	*Income group (%)*	*Consumption based CO2e emissions per capita 2013*
Richest	1	147.4
	9	31.7
	10	18.9
	10	15.6
	10	13.5
	10	11.8
	10	10.5
	10	9.2
	10	8.0
	10	6.4
Poorest	10	4.1

Source: Chancel and Piketty, 2015.

1.2 Country-specific studies

There are several national studies that attempt to put a figure on the pollution of the rich and compare this to the rest of the population. Jorgenson et al concluded that there were higher state-level emissions in US states where the wealthy captured more income. They found that combined consumption-based emissions from all sectors at the US federal state level were positively linked with the income share of the top 10%, top 5% and top 1% between 1997 and 2012.[9] This builds on a prior preliminary study by Jorgenson et al which found that income inequality (measured by the Theil household income inequality index) did increase residential carbon emissions in each state in the US.[10] Their hypothesis is that when income is concentrated at the top, there will be higher emissions because the rich will pollute via the overconsumption of goods and services (such as large homes, cars, boats and planes) which require high energy use. They also argue the rich use their political influence to evade measures to control emissions, an issue I cover in Chapter 4.

Ummel brought together data on the consumption of 6 million households between 2008 and 2012 in 52 areas of expenditure including electricity, gasoline, natural gas, heating oil, air travel, food and drink and other services. He estimated that the richest 10% of US polluters were responsible for around 25% of national emissions. He looked at the top 2% and found that their individual average footprint was over four times higher than those in the bottom 20% of the population. Crucially, for the top 2% around 75% of their emissions were from indirect sources, a finding that correlates with Chancel and Piketty's findings on the importance of indirect emissions.[11] Cohen built on Ummel's work to apply it to New York. He found that Manhattan, the richest borough, had the highest consumption-based emissions.[12]

In the UK there are a number of studies which find that the richest income group in the UK have high-consumption carbon footprints.[13] Druckman and Jackson found that the higher the disposable income the higher the carbon emissions. According to their study, the richest segment emitted 64% more carbon dioxide compared to the segment with the lowest emissions.[14] They based these findings on a quasi-multiregional input-output model which captured emissions from direct fuel use (personal vehicle use and flying) and in particular from emissions embedded in household goods and services. In earlier research they found that this trend deepened significantly between 1968 and 2000 because of high-income groups increasing their demand for travel (often by private car) and for energy resources such as heat and light in their homes.[15] In an earlier study Majima and Warde used

Family Expenditure Surveys in the UK between 1961 and 2004 to track the consumption patterns of the richest 1% by income. They noticed shifts during this time period from household goods to personal luxury items such as jewellery, a shift from luxury food and alcohol to eating out, increasing expenditure on domestic flights and expensive foreign holidays.[16] In summary, whilst recognising the diversity in lifestyle choices, I argue that *all rich individuals* in the US and the UK have a significant carbon footprint associated with their lifestyle.

1.3 Why do the richest consume in this way?

Veblen's theory of the leisure class from the early twentieth century is still a solid starting point in understanding why the richest people lead high-carbon luxury lifestyles.[17] Veblen argued that those who have accumulated wealth differentiate themselves from others through their leisure activities, which demonstrate they have the time to lead this lifestyle because they are able to "waste" time and effort on non-productive pursuits.

The basis of living in what Veblen calls "manifest ease and comfort" is possessing the monetary resources to pay for it. Leading this life of leisure requires sufficient wealth to purchase the tangible products which symbolise to others the different type of lifestyle this person can afford. Think of the use of private jets, they give a glimpse into the private lives of the rich because we can see that this is how they get around. But we do not know what happens in their lives in detail. The use of the private jet is enough to indicate that this person has the ability to lead a glamorous lifestyle.

Veblen then went on to observe that as wealth accumulates the "leisure class develops further in function and structure, and there arises a differentiation within the class" based on an elaborate system of rank and grades (including those who inherited their wealth and those who made it). This is where rich people's strategy of purchasing and using certain products to differentiate themselves from others (what Veblen called conspicuous consumption) takes on greater importance. Conspicuous consumption is a way of showing off how much wealth someone has. What is interesting is Veblen's observation that a fundamental factor was comparing what one had to what one's peers possessed. It was not enough to possess symbols of wealth, they had to be shown off to friends and competitors too through the "giving of valuable presents and expensive feasts and entertainments". A contemporary example in the UK would be the conspicuous consumption of Lord and Lady Bamford, which illustrates the environmental cost of luxury lifestyles. They have a record of flying guests to their parties on their private plane, by helicopter or by hiring planes. For example, in March 2016 they reportedly chartered two 737 jets to fly 180 friends to India to celebrate their 70th birthdays for four days in and around the palaces in Rajasthan.[18]

Since Veblen produced his seminal work on the leisure class there have been more recent studies charting the luxury lifestyle of the richest. For instance, Frank chronicles the inner workings of our divided society in what he calls *Richistan* and argues this type of consumption is status-seeking.[19] In another current study, Graham highlights the latest essential item for the rich: superyachts. Phil Popham, the CEO of one of the leading superyacht makers, observed in an interview that "People don't need yachts - they want yachts".[20] The competition between billionaires is possession of ever-larger superyachts (reaching nearly 600 feet in length). Being able to use a superyacht is a defining characteristic which demonstrates whether someone has elite status or not. As these luxury ways to travel are so expensive to purchase, maintain and rent they exclude the majority of people. For example, the average superyacht (around 47 metres in length) costs in the region of 30 million Euros and has annual maintenance costs of millions of Euros because of the crew and operation costs.[21]

1.3.1 Elite hypermobility as an expression of status consumption

The academic literature on elite consumption and elite mobilities tends to concur that the richest purchase, and also rent, luxury forms of transport as status symbols to prove they are members of the super-rich. These forms of mobility are taken for granted which locks rich people into a high-carbon lifestyle because not to travel in this way would mean that their membership of this elite class could be questioned. For example, the US President Donald Trump lives a high-carbon lifestyle, which includes his owning a fleet of aircraft (one of which is a Boeing 757) valued at $58 million.[22]

The richest people reconfirm their own identity as wealthy individuals through heavily polluting practices such as flying. Compared to poorer citizens who cannot afford to fly frequently they have more of their pride invested in a lifestyle that they could potentially lose if measures were implemented to restrict greenhouse gas emissions.[23]

Being hypermobile is a key way to demonstrate membership of the global elite. Multiple properties, luxury cars, private jets and yachts are the basis of the hypermobile lifestyle that many of the richest people lead. The point is these individuals are showing that they hold enough wealth to lead this lifestyle permanently. These are not one-off special occasions.

Having said that there are differences within the richest group. The level of hypermobility is an expression of how rich you are. Beaverstock and Faulconbridge[24] argue that at the *global* level there are:

1. The Ultra High Net Worth (UHNW) individuals (those with a net worth of US$30 million and above) who generally own a fleet of luxury planes, yachts, cars and helicopters. The latest Wealth-X report estimates that in 2017 there were 79,595 UHNW individuals in the US and 9,370 UHNW individuals in the UK.[25]
2. Below them the group with US$5 million net worth also frequently fly in private jets and use yachts but they are more likely to lease them. They are likely to own their own luxury cars though.
3. Finally, those with US$1 million net worth often travel first class and occasionally charter luxury transport. They also own their own luxury cars.

1.3.2 Hypermobility contributes to climate change

In a globalised world wealthy people have a strong desire to be hypermobile for personal and business reasons.[26] The richest have access to a global lifestyle; they are not restricted to a single place or country.[27] Elite mobilities lead to a high-carbon transport system involving more and more flights and driving which demands more runways and roads. Birtchnell and Caletrío argue that even though there is scientific consensus to address climate change there is a lack of action because this would conflict with the "mobility-as-usual" of the elite which is "stratified, intensively global, high carbon and expansionist".[28] Often the destinations which rich people are travelling to are also exclusionary (e.g. gated communities). Overall, this means that elite mobilities influence visions of "the good life" by promoting high-carbon luxury travel.

More wealth leads to higher volumes of consumption and therefore a higher carbon footprint. Transport is a central part of the carbon footprint of the richest for two reasons. Firstly, it is a key source of direct emissions (e.g. flying, driving, sailing a yacht). Secondly, as the richest consume more in volume, transport is a key source of indirect emissions which are embedded in the goods and services they use. The very richest tend to own a superyacht, as well as a private jet if they can afford one (some of them are increasingly renting private jets), and a collection of luxury cars. While they can only drive one of these cars at a time the indirect emissions are still embedded in the total number of cars they own. The materials and energy, and therefore emissions, used to build the cars are embedded in the car regardless of whether it is used or not. The embedded emissions in the possessions of the richest, especially when they do not even use them, could be considered wasteful consumption as these materials, time and energy could have been used for something else.[29]

1.3.3 Extreme overconsumption

The great wealth amassed by the richest means they have the opportunity to consume in more extreme ways that will damage the environment. For example, in future there is expected to be growing demand from the richest people to own and hire supersonic jets as part of their luxury lifestyles. Several of these jets are already being developed in the US and could begin flights in the early 2020s. They could cost between $100 million and $200 million each. While they are expected to be more fuel efficient than the Concorde, they will still emit around the same as a current plane.

Several companies already sell submarines known in the industry as "private submersible yachts".[30] The company Migaloo makes a range of submarines and is also developing a private island which they describe as a "private floating habitat based on existing semi-submersible platforms". Planned features include a penthouse 80 metres above sea level, a helicopter pad, spa and beauty salons, pool areas with waterfalls, a beach gym, outdoor cinema and a shark-feeding area.

When the richest people have significant emissions associated with their luxury lifestyles it is appropriate to ask ethical questions about this. Reflecting on the unequal distribution of emission between countries, Shue argued that the "central point about equity is that it is not equitable to ask some people to surrender necessities so that other people can retain luxuries".[31] I ask the reader to consider this statement in light of the differences in personal consumption emissions within countries. For countries such as the US and UK there are very likely to be huge disparities in consumption emissions between the richest 1% by income and the rest of the people. The comparison Shue makes does not directly apply to the US and UK because the majority of the population *are* able to meet their needs (although rising inequality means the poorest are increasingly relying on food banks and other forms of support). Nevertheless, it is still valid to consider his point about luxury emissions because of the considerable size of rich people's personal emissions.

Elite hypermobility and its consequent high-carbon lifestyle are constantly reinforced by the private wealth management companies, yacht and jet-operating groups and other service providers who devise, maintain, promote and give meaning to this way of life. As the owners, shareholders and executives of the manufacturers of private jets, cars, yachts and submarines the richest also profit from the increasing use of these high-carbon goods.

1.4 Implications

Whilst there is growing recognition of the responsibility of the richest for global warming, regardless of where they live, this has not yet been translated into public policy.[32] More research is needed to identify and test policies that could reduce the consumption carbon footprint of the richest.[33] If governments were to put such policies in place, a range of challenges would need to be taken into consideration. These include:

- The competition for conspicuous consumption means the richest might ignore policies that aim to get them to reduce their ecological footprint

Instead of simply trying to "keep up with the Joneses", the richest are in a different league from the rest of the population, competing over high-carbon lifestyles in areas such as the size of their superyachts.

- Environmental taxes may have less effect on the richest because they can afford to pay to continue polluting now, and in the future

There is little research on whether environmental taxes have directly led to the richest reducing their consumption of fossil fuels. What we need to know is can the richest afford to pay for higher taxes and continue to have a larger carbon footprint? Piketty has shown the rate of return on wealth is rising faster than the rate of economic growth.[34] Does this mean the richest, who hold a huge chunk of this wealth, could potentially use the returns generated from their existing wealth to cover the additional costs of environmental taxes indefinitely?

- The richest might not respond to initiatives that inform them about the ecological crisis and the damage their consumption does to the environment

A great deal of effort is going into making consumption more sustainable across all sectors of society, based on the assumption that if individuals have more information about the negative impact of their consumption then they will change their behaviour. But is it realistic to assume that informing the richest people of the negative environmental impact of building a larger mansion or flying in a private jet will result in changing their behaviour, when they are more concerned with competing over the size of their superyachts? Given the urgency of tackling the ecological crisis, is it dangerous to rely on the richest to voluntarily reduce their overconsumption?

I raise these questions because they are the type of difficult question that policymakers would have to engage with if they were to implement policies to reduce overconsumption by the richest people. However, I do not attempt to answer them here. Instead, later sections of this book (in particular Chapters 2 and 4) look at the role of the richest in shaping the consumption options available to everyone (including themselves) and how these options are dependent on fossil fuels. This distinguishes the richest people from the rest of the population and is the reason why this book focuses on their role in particular.

1.5 Conclusion

Chapter 1 found that, per person, the richest 1% by income in the US and the UK consume in more high-carbon ways compared to other citizens. More wealth leads to higher volumes of consumption (based on higher frequency of use) and therefore a higher carbon footprint. Given this higher consumption, there needs to be more focus on the richest to ensure their high-carbon lifestyles are not excluded from academic and public debates on consumption and inequalities.

Notes

1 DeCarlo, S., 2013, 18 September. Cost of Living Extremely Well Index: The Price of the Good Life. *Forbes*. Retrieved from: www.forbes.com/sites/scottdecarlo/2013/09/18/cost-of-living-extremely-well-index-the-price-of-the-good-life/#764b6c3d731b [Accessed 20 October 2017].
2 *Financial Times*. How to Spend It. Retrieved from: https://howtospendit.ft.com/ [Accessed 20 November 2018].
3 Berthe, A. and Elie, L., 2015. Mechanisms Explaining the Impact of Economic Inequality on Environmental Deterioration. *Ecological Economics*, [e-journal] 116 (August 2015). Retrieved from: www.sciencedirect.com/science/article/pii/S0921800915002116 [Accessed 4 May 2017].
4 Chancel, L. and Piketty, T., 2015. *Carbon and Inequality: From Kyoto to Paris. Paris School of Economics Working Paper*. Paris: Paris School of Economics.
5 Osuoka, I. and Zalik, A., 2016. No Change There! Wealth and Oil. *In*: Hay, I. and Beaverstock, J., eds. *Handbook on Wealth and the Super-Rich*. Cheltenham: Elgar, 455.
6 Gore, T., 2015. Extreme Carbon Inequality. *Oxfam International*. Retrieved from: www.oxfam.org/en/research/extreme-carbon-inequality [Accessed 6 October 2018].

7 Hubacek, K. et al., 2017. Global Carbon Inequality. *Energy, Ecology and Environment*, [online], 2 (6). Retrieved from: https://link.springer.com/article/10.1007/s40974-017-0072-9 [Accessed 4 December 2018].

8 Otto, L. et al., 2019. Shift the Focus from the Super-poor to the Super-rich. *Nature Climate Change*, [online], 9 (82–84). Retrieved from: www.nature.com/articles/s41558-019-0402-3 [Accessed 17 February 2019].

9 Jorgenson, A., Schor, J. and Huang, X., 2017. Income Inequality and Carbon Emissions in the United States: A State-level Analysis, 1997–2012. *Ecological Economics*, [online], 134 (40–48). Retrieved from: https://ideas.repec.org/a/eee/ecolec/v134y2017icp40-48.html [Accessed 17 February 2019].

10 Jorgenson, A. et al., 2015. Income Inequality and Residential Carbon Emissions in the United States: A Preliminary Analysis. *Human Ecology Review*, [online], 22 (1), 95. Retrieved from: www.jstor.org/stable/24875150?seq=1#page_scan_tab_contents [Accessed 4 December 2018].

11 Ummel, K., 2014. Who Pollutes? A Household-Level Database of America's Greenhouse Gas Footprint. Center for Global Development Working Paper 381. Retrieved from: www.cgdev.org/sites/default/files/who-pollutes-database-greenhouse-gas-footprint.pdf [Accessed 20 December 2018].

12 Cohen, D. A., 2016. Petro Gotham, People's Gotham. *In*: Solnit, R. and Jelly-Shapiro, J., eds. *Nonstop Metropolis: A New York Atlas*. Berkeley: University of California Press, 47–54.

13 Büchs, M. and Schnepf, S., 2013. Who Emits Most? Associations between Socio-economic Factors and UK Households' Home Energy, Transport, Indirect and Total CO2 Emissions. *Ecological Economics*, [online], 90 (June 2013). Retrieved from: doi:10.1016/j.ecolecon.2013.03.007 [Accessed 16 August 2018].

Gough, I. et al., 2011. The Distribution of Total Greenhouse Gas Emissions by Households in the UK, and Some Implications for Social Policy. *New Economics Foundation*, [online]. Retrieved from: http://b.3cdn.net/nefoundation/ff3ed7d482cf03b852_lwm6b4r3a.pdf [Accessed 16 October 2017].

14 Druckman, A. and Jackson, T., 2009. The Carbon Footprint of UK Households 1990–2004: A Socio-economically Disaggregated, Quasi-multi-regional Input–output Model. *Ecological Economics*, [online], 68 (7). Retrieved from: https://doi-org.proxy-lib.anglia.ac.uk/10.1016/j.ecolecon.2009.01.013 [Accessed 17 February 2018].

15 Druckman, A. and Jackson, T., 2008. Measuring Resource Inequalities: The Concepts and Methodology for an Area-based Gini Coefficient. *Ecological Economics*, [online], 62 (2). Retrieved from www.sciencedirect.com/science/article/pii/S0921800907006106 [Accessed 17 February 2018].

16 Mijima, S. and Warde, A., 2008. Elite Consumption in Britain 19621–2004: Results of a Preliminary Investigation. *In*: Savage, M. and Williams, K., eds. *Remembering Elites*. Oxford: Wiley-Blackwell, 235.

17 Veblen, T., 1924. *Theory of the Leisure Class*. London: Allen & Unwin.

Frank, R. H., 1999. *Luxury Fever: Weighing the Cost of Excess*. Princeton: Princeton University Press.

18 Ostler, C., 2016, 16 April. Britain's Blingest Couple. *The Daily Mail*. Retrieved from: www.pressreader.com/uk/scottish-daily-mail/20160416/281973196822493 [Accessed 20 November 2017].

19 Frank, R., 2008. *Richistan: A Journey through the American Wealth Boom and the Lives of the New Rich*. New York: Penguin Books.

20 Graham, E., 2016, 21 January. More Billionaires, More Superyachts by 2020, Says Sunseeker CEO. *Bloomberg*. Retrieved from: www.bloomberg.com/news/articles/2016-01-21/more-billionaires-more-superyachts-by-2020-says-sunseeker-ceo [Accessed 20 November 2017].
21 Spence, E., 2016. Eye-spy Wealth: Cultural Capital and 'Knowing Luxury' in the Identification of and Engagement with the Superrich. *Annals of Leisure Research*, [online], 19 (3). Retrieved from: doi:10.1080/11745398.2015.1122536 [Accessed 17 February 2018].
22 Gale, R., 2017, 29 June. Calculating Donald Trump's Personal Impact on the Environment. *Marie Claire*. Retrieved from: www.marieclaire.com/politics/news/a27836/donald-trump-carbon-footprint/ [Accessed 20 November 2017].
23 Malm, A., 2016. *Fossil Capital: The Rise of Steam Power and the Roots of Global Warming*. London: Verso.
24 Beaverstock, J. V. and Faulconbridge, J. R., 2014. Wealth Segmentation and the Mobilities of the Super-Rich: A Conceptual Framework. *In*: Birtchnell, T. and Caletrío, J., eds. *Elite Mobilities*. Oxford: Routledge, 47–50, 54.
25 Elkins, K., 2018, 18 September. 71 Percent of the World's Super Rich Live in These Ten Countries. *CNBC*. Retrieved from: www.cnbc.com/2018/09/17/71-percent-of-the-worlds-super-rich-live-in-these-10-countries.html [Accessed 20 November 2017].
26 Budd, L., 2016. Flights of Indulgence (Or How the Very Wealthy Fly): The Aeromobile Patterns and Practices of the Super-Rich. *In*: Hay, I. and Beaverstock, J., eds. *Handbook on Wealth and the Super-Rich*. Cheltenham: Elgar, 310–312; Elliot, A., Elsewhere: Tracking the Mobile Lives of Globals. *In*: Birtchnell, T. and Caletrío, J., eds. *Elite Mobilities*. Oxford: Routledge, 27–30.
27 Short, J. R., 2013. Economic Wealth and Political Power in the Second Gilded Age. *In*: Hay, I., eds. *Geographies of the Super-Rich*. Cheltenham: Elgar, 34.
28 Birtchnell, T. and Caletrío, J., 2014. Introduction: The Movement of the Few. *In*: Birtchnell, T. and Caletrío, J., eds. *Elite Mobilities*. Oxford: Routledge, 4–5.
29 Sayer, A., 2016. *Why We Can't Afford the Rich*. Bristol: Policy Press.
30 Abrams, M., 2016, 11 May. How the Rich Live Now: Supersonic Jets and Submersible Yachts. *Observer*. Retrieved from: https://observer.com/2016/05/how-the-rich-live-now-supersonic-jets-and-submersible-yachts/ [Accessed 20 November 2017].
31 Shue, H., 2010. Subsistence Emissions and Luxury Emissions. *In*: Gardiner, S. et al., eds. *Climate Ethics*. Oxford: Oxford University Press, 211–212.
32 Gough, I., 2017. *Heat, Greed and Human Need: Climate Change, Capitalism and Sustainable Wellbeing*. Cheltenham: Elgar.
33 Kenner, D., 2015. Inequality of Overconsumption: The Ecological Footprint of the Richest. GSI Working Paper 2015/2.
34 Piketty, T., 2014. *Capital in the Twenty-First Century*. Cambridge, MA: Harvard University Press.

2 The investment emissions of the polluter elite

High-carbon consumption is not unique to the wealthiest as many other people in the US and UK also drive vehicles and take flights. That said, the richest people have a larger carbon footprint per person. As Chapter 1 showed this is because the volume of high-carbon consumption is higher for the richest. As I began to look at the size of the emissions associated with shareholdings in polluting companies (e.g. extracting oil, gas and coal, or producing steel, cement and petrochemicals) I realised that this did differentiate wealthy individuals who owned these shares from the rest of the population. This is an important observation because to date when the richest have been discussed in relation to climate change in academic literature or popular culture it has often been in connection with their use of private luxury transport such as jets and yachts.

For many people climate change is abstract and difficult to comprehend. This is because the location of the sources of greenhouse gas emissions can appear disconnected from the location of the consequences (extreme weather events which are increasing in frequency and intensity). However, we do know that the key cause of climate change is emissions, in particular from the burning of fossil fuels (see Introduction). I argue that even though the decision to emit one tonne of emissions cannot always be directly connected to a specific extreme weather event this does not mean it is impossible to identify responsibility. Chapter 2 looks at the investment component of the *unequal ability to pollute*. Rich people who hold significant amounts of shares in polluting companies are in a markedly different position from the rest of the population. I call this group the *polluter elite*. I argue that they have a different level of personal historical responsibility for climate change. The *polluter elite* database published online alongside this book lists individuals with shares in polluting companies and connects the number of shares held to the percentage of the company's annual emissions. By focusing in on those individuals who profit from their shares in polluting companies it is possible to identify and begin to quantify responsibility and complicity.

2.1 Who are the *polluter elite*?

Why do I focus on the investments of the richest in polluting companies? Because they are much more likely to hold such investments than the rest of the population. The richest 1% hold most of the productive and financial assets, with real estate making up a smaller portion of their portfolios.[1] This is part of a broader trend where the types of capital have changed in importance over time from land to real estate to business assets. For the richest 9% who come below the richest 1%, the main asset is their primary residence. This section of the population could hold productive investments, but they would likely be smaller in size than those held by the richest 1% because of the sheer concentration of wealth at the top.

Below the richest 10% of the population, if citizens do hold wealth, it is likely to be the property they live in. The remainder of the population, in particular, the poorest, could hold no investments at all, and in fact, are more likely to be living stressful lives as they probably have to deal with debts (Figure 2.1).

There are two main ways in which a rich individual could hold these shares:

- **Directly**: These rich individuals could have purchased these shares in an oil company or received them as an employee or director of the company. The main focus in this book and the polluter database is on rich people who have direct shareholdings.

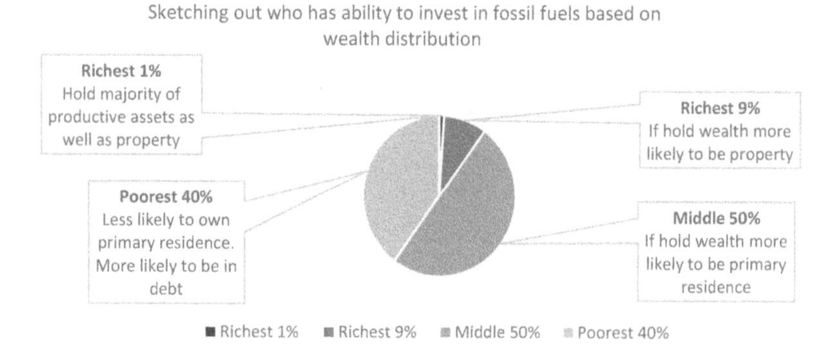

Figure 2.1 Which people are more likely to hold shares in polluting companies?
Source: Author.

- **Indirectly**: It is possible to profit from indirect investments made via asset managers (such as investment banks, private equity and hedge funds) which could be invested in polluting companies or directly in commodities such as oil and gas. This could have happened because of inaction; for example, the investor did not ask for their portfolio to be screened for oil, gas and coal companies.

I am not focusing on investments in these polluting companies via a pension fund because I consider this to be several steps removed from the decision of where to invest and this money is only received on retirement. My focus is on wealthy people who hold these shares as individuals and receive income from them throughout their lives.

Not all of the richest people in the US and the UK hold large investments in oil, gas, coal, and other major polluting industries. Some wealthy individuals invest in renewable energy because it is profitable. Often these rich individuals also publicly state they care about addressing climate change. For example, in the US a number of billionaires have investments in clean technologies (renewable energy and electric vehicles) including Tom Steyer, Jeremy Grantham, Andre Heinz, Nat Simons, Pierre Omidyar, Nicholas Pritzker, Greg Penner and Dick DeVos.

While other types of investments are not dealt with in detail here it is important to keep in mind that almost all investments will lead to an increase in economic activity which in turn will lead to higher emissions from production and distribution, often as part of complex global supply chains. The wealth of the richest people in the US, UK and across the world rests on economic growth which in its current form increases environmental degradation.[2] The current type of economic growth is pushing past the limits of the planetary boundaries leading Rockström to conclude "it is time to re-evaluate our economic and political models for the Anthropocene".[3] Planetary boundary thresholds already crossed include climate change, loss of biosphere integrity, land-system change and altered biogeochemical cycles.[4]

The focus of this book is on a group I call the *polluter elite*. These are the decision-makers at polluting companies such as the executive team (e.g. Chief Executive Officer, Chief Financial Officer, Vice Presidents) and directors. I use three criteria to identify the *polluter elite*:

1. These rich individuals have a net worth of at least $1 million, excluding their primary residence.
2. These are rich individuals with shares in a company that produces significant greenhouse gas emissions such as oil, gas, coal, steel, cement, chemicals and petrochemicals. The focus of this book is on oil and gas companies.

3 These rich individuals have the ability to shape the consumption choices of the general population to skew them towards lifestyles that are intertwined with fossil fuels so that the average citizen remains "addicted" to fossil fuels. This is as decision-makers at oil and gas companies who approve increasing extraction of fossil fuels and lobbying of political parties.

It was Mills who identified that the power elite in the US (who ran the interlocking political institutions, the military and the corporations) could "make decisions having major consequences" for ordinary men and women.[5] As Pirani reflects on fossil fuel consumption: "households are unlikely to make drastic changes to their consumption habits that would make a noticeable difference, without equally drastic changes in the economic, social and cultural contexts in which they live".[6]

2.2 The *polluter elite* database

One way to make the links between a rich person and their investments in companies that produce emissions is to look at how billionaires made their money. Based on the Forbes billionaire list an Oxfam International report estimates that globally the number of billionaires who have fossil fuel investments grew from 54 in 2010 to reach 88 in 2015. Over the same period their personal wealth rose from around $200 billion to $300 billion, a rise of nearly 50%.[7]

In 2016 Forbes identified US billionaires who had made their money from oil, gas, coal and infrastructure such as pipelines.[8] These included David and Charles Koch with a range of fossil fuel interests (discussed in greater detail in Chapter 4), and Warren Buffet who had $4.5 billion of shares in the oil refinery company Phillips 66, as well as controlling the rail freight company Burlington Northern Santa Fe Corp. which transports oil and coal. Buffet has also reportedly invested $30 billion in clean energy. Other examples include billionaires such as Randa Williams and family whose company Enterprise Products Partners owns 50,000 miles of pipelines to transport natural gas and oil.

At the top of the 2018 UK rich list was Sir Jim Ratcliffe, who had an estimated net worth of £21 billion. Ratcliffe is the founder and CEO of petrochemical giant INEOS, the largest private company in the UK which has licenses to do fracking. This is short for hydraulic fracturing which is a way to extract gas or oil trapped in shale rock by pumping large quantities of water, sand and chemicals down a deep well under the ground. Looking at the other top ten richest people in Britain, several of these billionaires' sources of wealth are linked to greenhouse gas emissions including Sri and Gopi Hinduja (oil and gas), Len Blavatnik (previously invested in Russian oil company TNK-BP), Lakshmi Mittal (steel) and Alisher Usmanov (steel).[9]

Whilst the lists of rich people are useful for discovering how the wealthiest people in the US and UK made their fortunes, it can be difficult to obtain accurate information about the annual greenhouse gas emissions of companies such as Koch Industries and the Hinduja Group as they do not disclose them. All that can be said is that they are likely to be huge. The Hinduja Group owns the Gulf oil company and has invested in a coal plant in India and in a company building vehicles (the emissions of Koch Industries are covered in Chapter 4).

It is also difficult to obtain accurate information on the investments in polluting companies of wealthy individuals in the US and the UK. Due to confidentiality surrounding rich people's investments it will not be possible in this book to estimate the total *investment emissions* of the richest 1% in the US and the UK who make up the *polluter elite*.

Even if it is known they do hold these investments, as the rich lists demonstrate, it is necessary to look at other sources of information to gain a broader picture. According to the Carbon Majors Database, based on work by Heede,[10] 100 companies are responsible for an estimated 71% of direct emissions and indirect emissions from extraction and burning of fossil fuels by industry and consumers between 1988 and 2015; of these 43 were state-owned or state producers. There were 57 private companies (41 were public investor-owned companies and another 16 were private investor-owned companies). My starting point for identifying the *investment emissions* of wealthy US and UK citizens in the *polluter elite* database were these 57 companies.

There are good reasons to focus on the world's largest investor-owned fossil fuel producers. Many of these companies have ignored their own climate scientists to continue production and several have systematically tried to prevent government action (discussed in Chapter 4).[11] These investor-owned companies do not operate in isolation from state companies (who hold most of the fossil fuel reserves) and arguably help secure demand for state producers. They do this by strengthening the web of global production and distribution of fossil fuels that help to entrench fossil fuels globally.

Whilst this book focuses on privately owned companies, a future research agenda could also explore state-owned companies in the Carbon Majors database, identifying the individuals who personally profit from their activities (whether as shareholders and/or via salaries).

What is possible is to show that certain rich people are shareholders in fossil fuel companies. This is why accompanying this book is an original database which attempts to calculate the emissions of the shareholders in these companies. The key source for this endeavour is the company annual report. The task is to put names and faces to the rich investors who, through their investments, are contributing to climate change. This is why the database also lists the major shareholders from the financial sector, such as large asset managers often based in the US and the UK, as this is another way that a rich individual could be indirectly invested in these companies.

This database compiles information on several hundred wealthy individuals with shareholdings in large oil and gas companies with historically significant emissions. This gives an initial insight into the *polluter elite*. As the database is based on publicly available information most of the shareholders listed in the database are members of the executive team (e.g. the CEO and CFO) and directors.

2.2.1 Methodology

There is currently no universally accepted way to calculate the greenhouse gas emissions of shares. Indeed, the objective of the *polluter elite* database is to establish a baseline which can be further improved to obtain increased levels of accuracy. The full database can be viewed online at https://whygreeneconomy.org/

There is a growing interest in measuring portfolio carbon footprints which is "the sum of a proportional amount of each portfolio company's emissions (proportional to the amount of stock held in the portfolio)".[12]

This has particularly been the case since the Financial Stability Board's high-profile Task Force on Climate-related Financial Disclosures was set up in 2015. In 2015 the French government passed legislation requiring French institutional investors to disclose the carbon risks of their portfolios. There is an increasing number of asset owners and managers signing the Montreal Pledge, which commits signatories to footprinting their portfolios. Investors want to understand the risks and opportunities.

There is a range of methodologies depending on the type of investment that is being measured.[13] In this book I will use the ASN Bank methodology for equity investments.[14] They give the following example for Nike:

> ASN Bank has around 190,000 shares in Nike. The value of these shares is around 15,000,000 Euro. This represents about 0.00022% of Nike's total Market Capitalisation. Based on Trucost

information, Nike has annual total carbon emissions of around 1,800,000 tonnes CO2eq. If we hold ASN Bank accountable for the proportional share of Nike's emissions, this results in indirect emissions of 396 tonnes CO2eq.

Taking a look at the company website and the documents prepared for the Annual General Meeting, filed with the Securities and Exchange Commission (for the US) and Companies House (for the UK) it is possible to see the names of the senior executive team and board members who are shareholders.

I apply the ASN Bank methodology to the individuals that I am able to identify as shareholders in the investor-owned oil and gas companies in the Carbon Majors Database. Many of these companies are based in the US. As a result, I have complemented this list by looking at shareholders (executive team and directors) in the biggest UK-based oil and gas companies, many of who have operations in the North Sea as part of their portfolio.

As the last date of available data from the Carbon Majors Database is for the year 2015, I have chosen to calculate the *investment emissions* of shareholders on the 31 December 2015 or as close as possible to this date. I had to pick a specific date because the share price and the number of shares can change. However, as companies report at different times it was not always possible to obtain information on shareholders as on 31 December 2015. For example, some oil and gas companies submitted their AGM information in March or April 2016 with information on shareholdings related to February 2016. In these cases, I have assumed the shareholding was the same as it was on 31 December 2015.

Case study: Rex Tillerson *investment emissions* from shareholding in ExxonMobil

ExxonMobil is one of the largest fossil fuel companies in the world with historic responsibility for emissions going back to the 1850s. According to the Carbon Majors database, between 1988 and 2015 its estimated cumulative emissions were 17,785 million metric tons of CO2e. Using this methodology, it is possible to say that Rex Tillerson is responsible for 23,212 metric tons of CO2e for the year 2015 (Figure 2.2). For the first three

(*Continued*)

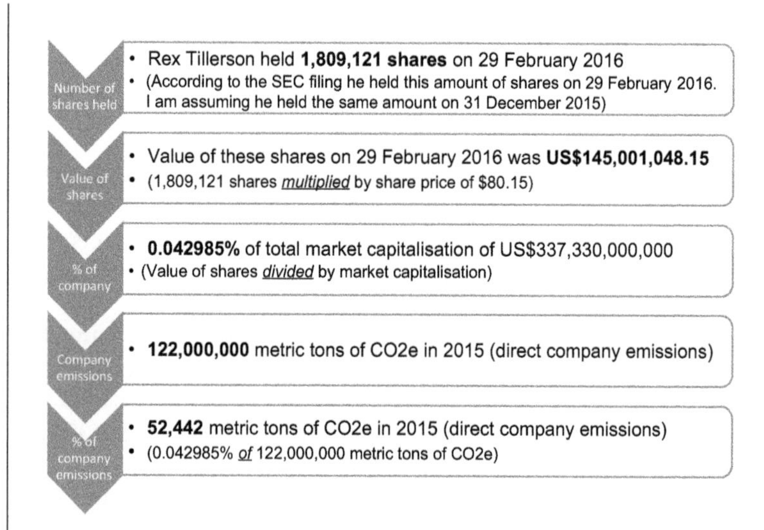

Figure 2.2 Rex Tillerson – Estimated investment emissions for shareholding in ExxonMobil in 2015.

sections below, the source is the ExxonMobil filing of its Definitive Proxy Statement Form 14A to the SEC.[15] The company emissions are those self-disclosed by ExxonMobil. The percentage of company emissions are my own calculations of *investment emissions*.

Case study: Bob Dudley *investment emissions* from shareholding in BP

BP is one of the largest fossil fuel companies in the world with historic responsibility for emissions. According to the Carbon Majors Database, between 1988 and 2015 its estimated cumulative emissions were 13,791 million metric tons of CO2e. For the first three sections below, the source is the BP annual report 2015.[16] The company emissions are those self-disclosed by BP. The % of company emissions are my own calculations of *investment emissions*. Using this methodology, it is possible to say that Bob Dudley is responsible for 2,292 metric tons of CO2e for the year 2015 (Figure 2.3).

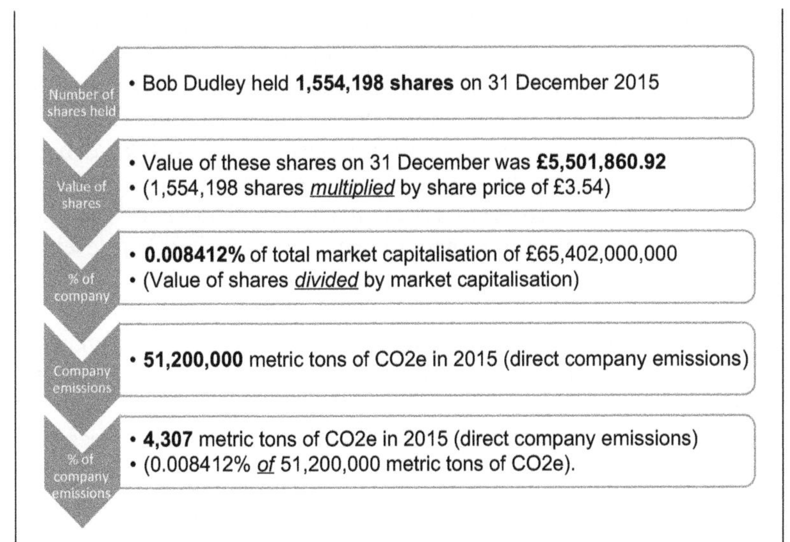

Figure 2.3 Bob Dudley – Estimated investment emissions for shareholding in BP in 2015.

The *polluter elite* database also captures profits from shareholdings in a polluting company. For example, continuing with the case of Bob Dudley, he held 1,554,198 shares on 31 December 2015 (worth £5,501,860.92). Based on the dividend of £0.07 for the fourth quarter (October – December 2015) he received £108,988.13. This dividend was announced on the 2 February 2016 and paid out on the 24 March 2016.[17] Note that Dudley would have also received around £300,000 more from dividends for the other three quarters which paid out on 19 June, 18 September and 18 December 2015. This means that in 2015 Dudley would have received dividends totalling over £400,000. The *polluter elite* have several options on how to use these dividends. They could include funding a high-carbon lifestyle or in some cases making personal donations to political parties (see Chapter 4).

2.3 Discussion of findings from the *polluter elite* database

What company emissions are the *polluter elite* responsible for? The objective of this database is to show that it is possible to put a figure on the personal emissions of shareholders in large polluting companies.

As explained above I am applying one approach to calculate portfolio emissions connected to an individual's shareholding. As far as I am aware this has not been done before and therefore, I am sure there are ways to improve the methodology to obtain greater accuracy. My hope is that by publishing the *polluter elite* database this will spark a debate about how to connect a shareholder to total company emissions.

Before looking at different types of shareholders there are two common factors which unite them in terms of exploring their level of responsibility:

- Firstly, when they receive dividends, they are profiting from economic activity that causes pollution which they know is contributing to climate change.
- Secondly, the individuals who hold shares in these companies enable the company to operate. Their investment is used to fund operations, in particular exploration for new reserves. This enabling happens alongside other larger sources of finance for the company such as loans from banks and the bonds raised by the company.

I will begin a discussion of responsibility by focusing on the executive team and directors who are listed in the *polluter elite* database. I argue these individuals hold a high level of responsibility because, through their action of running the companies listed in the *polluter elite* database, they are consolidating and deepening the extraction of fossil fuels. It is the executive team and the directors who decide how the company will operate and therefore affect the level of pollution they cause. They choose whether the company invests in offshore oil extraction or invests in solar panels. Therefore, I argue they hold more responsibility for emissions. In the case of the executive team and directors they also have the highest responsibility because they make decisions on their companies' operations and approve lobbying to consolidate the fossil fuel economy. Instead, they could be using their position to push their company to invest in renewable energy and for the state to support this. The cases of the energy companies such as Equinor (previously Statoil), Engie and Ørsted (previously DONG) are symbolic because they have taken the strategic decision to begin to shift to renewable energy on a large scale.[18] These three examples show that the executive team and directors do have a choice. While several oil and gas companies are investing more in renewable energy[19] and other innovations (ExxonMobil is investing in algae biofuels)[20] these are still relatively small amounts in terms of total investment and production quantity.

For these reasons, I think it is appropriate to link the shareholdings of the executive team and directors to the company's direct emissions. The Carbon Majors study, which was the starting point for the *polluter elite* database, calculates direct emissions from owned or controlled sources.[21] When I discuss the *polluter elite*'s responsibility my focus is on direct company emissions. However, before proceeding it is important to briefly cover indirect emissions.[22] This is why I also include a figure on an individual's percentage of each company's indirect emissions along the supply chain.[23] According to the Carbon Majors study these indirect emissions account for 90% of total company emissions. It is debatable to what extent indirect emissions are applicable to the *polluter elite* because without the extraction by the fossil fuel companies nobody else, whether industry or citizen, could burn fossil fuels for their own purposes. Perhaps it is more accurate to say that the *polluter elite* hold responsibility for the extraction of fossil fuels, which they know full well will then be burnt, but that it is the final user who is responsible for the actual emissions. This is the argument often used by these companies who say they are simply supplying what consumers demand. However, this is not the whole story for two key reasons which Chapter 4 will cover in more detail. Firstly, these companies, and the *polluter elite* who run them, lobby the state to maintain the fossil fuel status quo which then determines the options available for end users of energy (industry or consumers). Secondly, these companies have known about climate change since around the 1970s meaning they knew the consequence of extracting fossil fuels would be a rise in direct and indirect emissions which would contribute to global warming.[24] Instead of changing their behaviour these companies, and the *polluter elite* who run them and profit from their shareholdings, have consistently decided to deepen extraction, find more reserves and lobby to entrench the fossil fuel dependence throughout the global economy, in order to return higher dividends to shareholders (which is their fiduciary duty of course).

By choosing fossil fuels ahead of alternative energy the decision-makers in these polluting companies have a particular responsibility because they are shaping the energy choices of the general population in many countries now and in the future. There is carbon lock-in whereby it is easier to continue with systems that are already running on fossil fuels. This is particularly the case for fossil-fuel based energy infrastructure because of large capital costs and the time (often several decades) this infrastructure can be used for. Continuing to build cars which use petrol and diesel is bound up with infrastructure lock-in for roads and the built environment.[25] This means that the main energy option for travel is based on oil (instead of electric cars), for heating a home is based on gas (instead of electric heat pumps and/or combined heat and power) and for electricity based on gas or coal (instead of renewable energy). Of course, many people in the US and the UK own cars that pollute the environment which affects the health of pedestrians, and other drivers including themselves. But it is the decision makers in polluting companies who have the power to influence the energy choices for the general public. Indeed, they do this on purpose to secure a market for their product. Therefore, is it right to place the onus on individual consumers to green their behaviour when the executive team and directors are working to maintain the fossil fuel status quo?[26] Without them, nobody else, whether industry or citizen, can burn fossil fuels for their own purposes.

2.3.1 Differing levels of complicity

The focus of this book is very much the decision makers as they are the ones that approve further extraction and lobbying. What about other shareholders who are not the decision makers in the company? I will argue that for other members of the *polluter elite* their level of complicity hinges on their level of agency in deciding to profit from pollution. While there is huge diversity in shareholders (indeed, some use their shareholding to push polluting companies to respond to the issue of climate change) I distinguish between two types of shareholders:

- A shareholder who actively decides to hold shares in a polluting company.

If a rich individual knows that a company is going to pollute (and how could they avoid knowing if it is an oil, gas or coal company) then they are complicit in the pollution. They actively and knowingly choose to profit from this polluting investment. They are aware that this human activity is leading to global warming but continue to invest in these companies. Although they do not hold as much responsibility as the directors, they are complicit in the extraction of fossil fuels. They are supporting the status quo of globalisation powered by fossil fuels for production and consumption. This maintains the web of fossil fuels and entrenches carbon lock-in.[27]

This would apply to shareholders who held substantial shares. Based on the calculations in the *polluter elite* database the following individuals were complicit with significant emissions in 2015: Carl Icahn in Chesapeake (504,428 metric tons of CO2e of direct company emissions) and Aristotelis Mistakidis in Glencore (761,654 metric tons of CO2e of direct company emissions).

Former executive team members and Directors are also implicated. For example, while a CEO may have left the company, they are likely to still hold a substantial chunk of shares which they were awarded whilst an employee of the company. While they are no longer a decision maker in that company, they have chosen to hold on to their shares and profit from pollution. This is less responsibility than a current decision maker at the company but when the company registers higher revenues they still profit.

- A shareholder who does not ask the asset manager to screen investments for fossil fuel companies and other major polluters.

These people have chosen not to apply exclusionary screens to fossil fuels. For example, it is already possible to avoid exposure to companies who operate in controversial sectors such as fossil fuels, tobacco or weapons by screening them out. This is less complicity but because of their inaction they are still invested in fossil fuels, and crucially, still profiting.

2.3.2 Historical responsibility for climate change

The above case studies of Rex Tillerson and Bob Dudley show that it is possible to have large personal emissions from shareholdings. For comparison in terms of consumption the highest emitting activity any individual can do is to fly. A long-haul flight such as a one-way flight from London to Los Angeles leads to an estimated 1.7–2.9 metric tons of CO2e according to online carbon footprint calculators. This is for *one* passenger in economy class (not business or first class) on a commercial plane travelling a distance of close to 8,800 km. For an individual to have a carbon footprint at the same level as the *investment emissions* of one of the CEOs mentioned above, they would have to take thousands of long-haul flights in a year.

Indeed, the differences in magnitude are likely to be even larger than those documented in the *polluter elite* database because these figures are just based on the shareholding an individual has in one company for the year 2015. The *polluter elite* database does not capture the *investment emissions* of these rich individuals (and indeed those who were in the executive team and held directorships before them who are not in the database) in the years before or after 2015. The database also does not calculate the total *investment emissions* for the year 2015. As I did the research, I noticed that the company annual report had information about other current or past positions at oil, gas and coal companies which would imply these people also held shares in those companies (i.e. directorships or executive positions). This is a window into other shareholdings in polluting companies and could be the starting point to begin to comprehend the total *investment emissions* of a rich individual at a given point, for example, 31 December 2015.

Therefore, based on the findings in the *polluter elite* database it is reasonable to conclude that when a rich individual in the US or the UK has significant *investment emissions* (like those decision makers listed in the *polluter elite* database) then they are more responsible for climate change compared to an individual whose personal emissions only come from their consumption. As I argued above (in section 2.1 "Who are the *polluter elite*?") this is the majority of the population because they do not hold productive investments in fossil fuels and other highly polluting companies.

The *polluter elite* database uses data from 2015 because this was the latest year for which data was available based on the Carbon Majors study. However, it is likely that the members of the *polluter elite* listed have *investment emissions* from previous years. When I was doing the research, I noticed in the company annual report that the profiles of the executive team and directors often said how long they had held this position. In many cases the rich individuals cited in the *polluter elite* database had been directors for at least ten years.

This matters because as mentioned above the Carbon Majors database found that 100 companies (private and state-owned) were responsible for an estimated 71% of direct and indirect emissions between 1988 and 2015. The period 1988–2015 is crucial to study because these 100 companies emitted more greenhouse gases during this time (estimated total 833 GtCO2e) compared to the years 1751–1988 (estimated total 820 GtCO2e).

Climate science research has made it clear that what matters are the total emissions that have accumulated in the atmosphere since the Industrial Revolution began in the UK. One additional tonne of carbon dioxide helps to tip the balance towards irreversible climate change. Therefore, it is pertinent to ask how long these individuals have held these shares and what their cumulative emissions are over time. By tracking the main shareholders of these companies, it would be possible to see which rich individuals from the US and UK, as well as other countries, have historically profited from pollution. Due to space limitations this is beyond the scope of this book.

BOX 2.1 A HISTORY OF PROFITING FROM FOSSIL FUELS: BRITISH COAL KINGS AND AMERICAN OIL MILLIONAIRES

The richest in the US and UK have a long track record of profiting from fossil fuels[28] and therefore have historical *investment emissions.* Prior to the coal industry being nationalised in 1947 many rich families in the UK made fortunes from coal investments.[29] To give some context, it is estimated that total profits from coal mining and related activities across the entire industry steadily rose from £1.66 million in 1849, reached £14.09 million in 1874, then fell until the 1890s, and by 1912 rose again to £19.96 million. Some of the key coal owners became very wealthy during their lifetimes. For example, John Straker left close to £1 million when he died in 1885; John Joicey left £710,495 in 1881 to his nephew James. By the time James Joicey died in 1936 he was Lord Joicey and his estate was worth approximately £1.5 million plus a further £400,000 in shares in coal collieries. David A. Thomas inherited £200,000 from his father's Welsh colliery business in 1879. He left £1,169,000 when he died in 1918. In an example of political influence, he was also the MP for Merthyr Tydfil between 1888 and 1906.

By 1850 some of the main coal owners were members of the aristocracy. Lord Durham was one of the country's largest coal producers until the 1890s. His income from collieries was around £84,000 in 1856, £122,658 in 1872 and £380,000 in 1873. Lord Durham later sold his collieries in 1913 but still received around £60,000 from wayleaves, royalties and railway rents from coal sources. Meanwhile, the Duke of Northumberland leased his land for coal production which led to an income of £82,450 in 1918.[30] The Charlesworth family were the owners of several collieries (Kilnhurst, Thrybergh, Rothwell Haigh, Robin Hood, Newmarket, Haigh Moor) which in 1889 were valued at £373,591. When J. C. D Charlesworth died in 1882 apart from his interests in his own coal pits, he had shares in Sheffield Steel and South Yorkshire Coal and also in the railways which transported the coal that his collieries produced.

In the US, the richest people have been investing in the economic activities that have produced greenhouse gas emissions for over a century. For example, during the period 1901–1914 some of the

(*Continued*)

wealthiest people derived their fortunes from oil and steel. Phillips documents that at this time the richest people in oil were John D. Rockefeller ($1 billion), Oliver Payne ($100–150 million), Henry Rogers ($100 million), William Rockefeller ($100 million), Henry Flagler ($75 million) and Charles Harkness ($75 million), and in steel, Andrew Carnegie ($400 million), Henry C. Frick ($150 million) and Henry Phipps ($75 million).[31] These same individuals and their descendants were still the top wealth holders by 1924 having been joined by automobile manufacturers such as the Ford family. There are connections between these historic investments and wealth today. Around 200 of Rockefeller's descendants who have inherited his fortune are believed to be worth around US$10 billion.

2.3.3 Beyond comparisons between countries

When the United Nations Framework Convention on Climate Change (UNFCCC) was signed in 1992 the richest countries agreed to take the lead given their historical responsibility for greenhouse gas emissions associated with their earlier process of industrialisation. Under Article 3 of the Convention, countries should participate "on the basis of equity and in accordance with their common but differentiated responsibilities and respective capabilities". Discussions at the UNFCCC negotiations revolve around comparisons between total emissions within a country's borders, with a focus on production emissions, for example, pollution from factories, fossil fuel power plants and the transport sector.

Strong arguments have been made by governments and civil society movements from poorer countries that the richest countries have a moral obligation to reduce emissions first in order to free up space for them to "grow". This is because richer countries have benefited the most from these emissions whilst poorer countries have suffered more of the consequences and have fewer resources to adapt. China has argued that as the largest historic global emitter the US should make more stringent emissions cuts. However, China became the world's largest emitter in the mid-2000s and it is now common to hear the US argue that China should not be classed as a "developing country" and should be forced to commit to ambitious emission reductions.

A related argument has been to compare per capita emissions between countries. This is where the total territorial emissions are divided by the total population. The key argument here has been that while emerging economies such as China and India have become the leading emitters, they also have huge populations of over a billion people, and therefore compared to countries such as the US or European countries they have lower emissions. For example, using this measure in 2014 the average Chinese citizen was responsible for 8.50 metric tons of CO2e compared to the average US citizen who emitted 19.84 metric tons.[32]

Another argument has been that focusing on emissions produced within the borders of a country ignores the fact that in the era of globalisation it is increasingly common for goods and services to be produced within several countries and then finally consumed in another country. There is a growing trend of production emissions, for example from power plants providing electricity or from a factory, being embodied in products that are then consumed in another country. Outsourced consumption-based emissions are increasingly significant for the US and the UK. The issue of outsourced emissions led to an evolution of the debate about which countries needed to do the most to reduce emissions. For example, even though China became the biggest global emitter in the mid-2000s it was starting to be acknowledged that the US had outsourced a lot of its manufacturing to China. According to a Deloitte study, outsourced emissions, which are counted as national consumption-based emissions, are a significant portion of total emissions.[33] In 2011 these made up an estimated 52% of national emissions in the UK and 22% in the US. Whilst the US has a lower percentage it is the largest net energy emissions importer in the G20 by quite some way. The main source of embedded emissions for both countries is China.

The UK has seen a fall in total emissions by 41.1% between 1990 and 2016, mainly due to the reduction of coal for electricity generation.[34] However, if emissions embedded in goods and services imported from other countries are factored in, the country has not seen any dramatic reduction in overall emissions as would appear at first glance from its domestic emissions.[35] For all these reasons it is important to compare the total greenhouse gas emissions of each nation to identify where unsustainable production and consumption patterns are concentrated globally. However, this has tended to hide the inequality of emissions within countries, something which this book aims to address.

Chakravarty et al proposed a focus on the one billion highest emitters regardless of where they lived as a way to get past the "developed/developing country" impasse at the climate talks.[36] They suggested spreading responsibility for reductions among individuals rather than countries which would mean every country participating. They found that the countries that needed reductions the most were the US and

China, followed by Russia, India, South Africa and countries in the Middle East. In an interview at the time with *Scientific American* one of the co-authors, Socolow, said, “rich people in poor countries shouldn’t be able to hide behind the poor people in those countries”.[37]

In their proposal for a Greenhouse Development Rights Framework, Baer et al defined capacity and responsibility of individuals in all countries for emissions. The proposal also factored in income inequality within countries, a move which the authors argued was long overdue.[38] A more recent follow-up study concluded that as part of the global effort to reduce emissions they should be divided fairly within countries so that the wealthiest do the most to bring down their carbon footprint.[39]

These existing debates have tended to focus on emissions related to consumption. The findings in the *polluter elite* database are a contribution to these debates. This book focuses on rich individuals from the US and the UK but as the database shows their investments are not restricted to the borders of the US and the UK. They invest in large multinational companies with international operations. They also invest in Chinese, Indian and Indonesian coal companies (to mention only a few) via the US- and UK-based financial intermediaries. When these people invest, they are doing so globally and therefore the pollution of the companies they invest in (often multinationals with global reach and part of global supply chains) is also global. This makes it harder to pin them down but what can be said is that the pollution from these rich individuals through their investments is global. Therefore, a discussion of their global responsibility and complicity is justified (Figure 2.4).

If the richest people in the US and the UK (and in many other countries around the world) are the ones who profit and hold responsibility and complicity for emissions produced in another country such as China, this justifies a focus on rich investors as key agents rather than the general population of the US or the UK and China. If it is wealthy US and UK citizens who profit from oil extraction in the Niger Delta, causing subsequent localised oil spills as well as contributing to global climate change, then arguably it is not accurate to count the greenhouse gas emissions from this oil extraction as part of Nigeria’s national emissions.

Figure 2.4 The investment emissions of the polluter elite transcend territorial borders.

Source: Author.

In terms of investments one of the key observations of the *polluter elite* database is that it captures the globalised nature of the economy. This matters in terms of who is responsible for pollution because, as the database documents, the *polluter elite* often invest in companies which operate globally outside of the national territory where they are citizens. For example, it is likely that wealthy individuals in the US and UK hold shares in Chinese and Indian coal companies via asset managers that are based in the US and the UK. While debates on ethical responsibility in relation to climate change have often focused on comparisons between high-emitting countries such as the US and China there is another dimension to this discussion which is *who* in both countries is responsible for these emissions happening?

If we trace private investments it becomes clear that nuance is needed when making an observation, such as that Western countries have overseen a process of deindustrialization and outsourcing of production to China which has dramatically increased that country's emissions (a huge chunk of which are for export). This statement does not capture the fact that there are rich citizens (often from North America or Europe) who still hold degrees of responsibility and complicity for these emissions whether this was pre-deindustrialization (pollution created in the West) or post-deindustrialization (pollution created in China). Therefore, to blame the entire Chinese population for national emissions going up is not just unfair but misleading because the actors facilitating the pollution, which most impacts Chinese citizens such as air and water pollution, by investing in these companies, remain hidden from view. In conclusion, in a globalised world tracking individual *investment emissions* is a more accurate way of identifying responsibility and complicity. This is more nuanced than comparing total or per capita production and consumption-based emissions between countries.

2.3.4 Recognising the role of the **polluter elite** *in the Anthropocene*

There is a growing consensus forming that we are now in the Anthropocene era where human activity has a dominant influence on geology and ecology[40] rather than the other way round. The Anthropocene has been presented in three broad phases.[41]

1 Since the beginning of the Industrial Revolution in the United Kingdom in the eighteenth century there has been an increase in the burning of fossil fuels.
2 The *Great Acceleration* followed the Second World War. This concept has been used to describe the profound impact of human

activity on the natural world since the 1950s. Indicators used to measure this process of change include socio-economic trends such as the rapid expansion of global GDP, transport and urbanisation alongside earth-system trends such as tropical forest loss and ocean acidification.

3 This culminated in growing awareness of the human role in environmental problems which was symbolised by the Rio Earth Summit in 1992.

While the extent to which we have entered the Anthropocene is still disputed it has become a rallying point for a range of actors such as ecological movements and scientists who are concerned about the impact of human activity. However, the Anthropocene narrative has been criticised for depoliticising the drivers of environmental degradation[42] and for distracting from the fact that a rich and powerful minority have created problems which have then been blamed on the majority.[43]

Steffen, one of the leading thinkers on the Anthropocene, has since clarified that humanity should not be treated as one homogenous whole because it is industrialised societies that have caused the Anthropocene, and not others such as indigenous people in Canada or small farmers in sub-Saharan Africa.[44]

To recognise the role of profit-seeking as the main historical driver of environmental degradation Malm argues the term "capitalocene" should replace the Anthropocene. He concludes, "this is the geology not of mankind, but of capital accumulation".[45] Meanwhile, Moore observes, "global warming is not the accomplishment of an abstract humanity, the *Anthropos*. Global warming is capital's crowning achievement".[46] From this perspective the Anthropocene should not just be seen as a scientific trend with a focus on human nature and an inevitable rise in population and resource use. To do so minimises the historic driver of the current ecological crisis which is the capitalist system.

There are a growing number of scholars who argue for a specific focus on the richest to reflect that the Anthropocene has been caused by a minority, implying different levels of responsibility. In recognition of the significant environmental damage caused by a wealthy minority Ulvila and Wilén have proposed to call the new epoch the "plutocene" (the era of the wealthy).[47] The aim is to re-politicise debates by identifying and challenging vested interests who are blocking a transition away from fossil fuels. They argue this is necessary because to date much debate has revolved around comparisons between countries which glosses over that the wealthy minority has more obligation to reduce their environmental impact compared to those below them.

While debates will continue on the relative merits of terms such as the "plutocene" and "capitalocene" the Anthropocene concept certainly has value because it captures the scale of human impact on the environment in a way that is gaining traction. Using the Anthropocene as the overall prism I argue that, based on the findings of the *polluter elite* database, a more detailed focus on the *polluter elite* is justified. This means moving from the macro (economic systems and theory) to the next level down, identifying the people who are driving ever-increasing emissions, particularly by using their wealth to secure political influence (dealt with in Chapter 4), and who personally profit. It is important to recognise the role of the *polluter elite* in causing climate change and in deepening the status quo.

2.4 Conclusion

The *polluter elite* database compiles information on several hundred wealthy individuals with shareholdings in the privately owned oil and gas companies with the largest historical emissions. They are some of the Carbon Majors identified by Heede (2014).[48] The database uses companies' annual reports to calculate the percentage of company emissions linked to publicly visible shareholders such as the CEO and Directors (where possible I also calculate the *investment emissions* of asset managers). These decision-makers form part of a group I call the *polluter elite*. I have argued that as *investment emissions* are so much larger than consumption emissions the *polluter elite* hold more responsibility for climate change. I have also reflected on what these findings mean for debates on climate change which compare emissions between countries.

Notes

1 Piketty, T., 2014. *Capital in the Twenty-First century*. Cambridge, MA: Harvard University Press, 68–69, 325.
2 Di Muzio, T., 2015. *The 1% and the Rest of Us: A Political Economy of Dominant Ownership*. London: Zed Books, 161–163.
3 Stockholm Resilience Centre. Planetary Boundaries Publications. Retrieved from: www.stockholmresilience.org/research/planetary-boundaries/planetary-boundaries/publications.html [Accessed 20 August 2018].
4 Potsdam Institute. Four of Nine Planetary Boundaries Now Crossed. Retrieved from: www.pik-potsdam.de/news/press-releases/four-of-nine-planetary-boundaries-now-crossed [Accessed 27 August 2018].
5 Mills, C. W., 1959. *The Power Elite*. New York: Oxford University Press.
6 Pirani, S., 2018. *Burning Up: A History of Fossil Fuel Consumption*. London: Verso Books.

7 Singer, M., 2018. *Climate Change and Social Inequality: The Health and Social Costs of Global Warming.* London: Routledge Books.
8 Helman, C., 2016, 9 March. The 93 Billionaires in Global Oil and Energy, 2016. *Forbes.* Retrieved from: http://fortune.com/2015/04/25/billionaires-versus-big-oil/ [Accessed 20 August 2017].
9 Jamieson, S., 2018, 14 May. Rich List: Jim Ratcliffe Named as UK's Richest Man and Record Number of Women among 1,000 Wealthiest. *The Daily Telegraph.* Retrieved from: www.telegraph.co.uk/news/2018/05/12/rich-list-record-number-women-among-1000-wealthiest-people-britain/ [Accessed 20 November 2018].
10 Heede, R., 2014. Tracing Anthropogenic Carbon dioxide and Methane Emissions to Fossil Fuel and Cement Producers, 1854–2010. *Climate Change,* [online], 122 (1–2). Retrieved from: https://link.springer.com/article/10.1007/s10584-013-0986-y [Accessed 17 February 2019]; Heede, R., 2017. Carbon Producers' tar pit: Dinosaurs Beware. *Climate Accountability Institute.* Retrieved from: www.ineteconomics.org/uploads/papers/Heede-PathToAccountability-18Oct17.pdf [Accessed 20 December 2018].
11 Frumhoff, P., Heede, R. and Oreskes, N., 2015. The Climate Responsibilities of Industrial Carbon Producers. *Climate Change,* [online], 132 (2). Retrieved from: https://link.springer.com/article/10.1007/s10584-015-1472-5 [Accessed 17 February 2018].
12 Principles for Responsible Investment. How Measuring a Portfolio Carbon Footprint Can Help. Retrieved from: www.unpri.org/climate-change/how-measuring-a-portfolio-carbon-footprint-can-assist-in-climate-risk-mitigation-and-reducing-emissions/608.article [Accessed 28 January 2019].
13 Rathbones Greenbank Investments. Portfolio Carbon Footprinting and Assessing Climate Risk. Retrieved from: www.rathbonegreenbank.com/insight/portfolio-carbon-footprinting-and-assessing-climate-risk [Accessed 28 January 2019].
14 ASN Bank, 2015. ASN Bank Carbon Profit and Loss Methodology. *ASN Bank.* Retrieved from: www.ecofys.com/files/files/4501704_asn_carbon-profit-and-loss-methodology-v5.pdf [Accessed 20 December 2018]; ASN Bank, 2017. ASN Bank Carbon Profit and Loss Methodology. *ASN Bank.* Retrieved from: www.asnbank.nl/web/file?uuid=a3a2d821-334e-4742-afa8-920a5d9ec61d&owner=6916ad14-918d-4ea8-80ac-f71f0ff1928e&contentid=765 [Accessed 20 December 2018].
15 ExxonMobil. ExxonMobil SEC filing, Definitive Proxy Statement Form 14A. Retrieved from: www.sec.gov/Archives/edgar/data/34088/000119312516539460/d14941ddef14a.htm [Accessed 20 August 2018].
16 BP. BP Annual Report 2015. Retrieved from: www.bp.com/content/dam/bp/business-sites/en/global/corporate/pdfs/investors/bp-annual-report-and-form-20f-2015.pdf [Accessed 20 August 2018].
17 BP. BP Cash Dividends – Ordinary Shareholders. Retrieved from: www.bp.com/content/dam/bp/business-sites/en/global/corporate/pdfs/investors/bp-cash-dividends-ordinary-shareholders.pdf [Accessed 20 August 2018].
18 Mäkitie, T. et al., 2018. The Green Flings: Market Fluctuations and Incumbent Energy Industries' Engagement in Renewable Energy. Working Papers on Innovation Studies 20180524, Centre for Technology, Innovation and Culture, University of Oslo. Retrieved from: https://ideas.repec.org/p/tik/inowpp/20180524.html [Accessed 17 February 2019].

19 BP. Alternative Energy. Retrieved from: www.bp.com/en/global/corporate/what-we-do/alternative-energy.html [Accessed 3 March 2019].
20 ExxonMobil. Advanced Biofuels. Retrieved from: https://corporate.exxonmobil.com/research-and-innovation/advanced-biofuels [Accessed 3 March 2019].
21 Where possible I have also included an estimate of investment emissions connected to emissions disclosed by the company. I did this because there can be significant differences compared to the direct emissions calculated in the Carbon Majors study. For example, for the year 2015 ExxonMobil disclosed total emissions of 122 million tonnes of CO2e, which includes their upstream, downstream and chemical operations. Whereas in the Carbon Majors study ExxonMobil's direct emissions were estimated to be 54 million tonnes of CO2e.
22 Greenhouse Gas Protocol. FAQ. Retrieved from: https://ghgprotocol.org/sites/default/files/standards_supporting/FAQ.pdf [Accessed 20 August 2018].
23 There are various ways the emissions associated with a company can be calculated. These include: Scope 1 emissions which are direct emissions from owned or controlled sources. These direct emissions are the focus in this book; Scope 2 emissions from the generation of purchased energy that are classified as indirect emissions; and finally Scope 3 emissions that are indirect emissions throughout the value chain from upstream (e.g. extraction) to downstream (e.g. burning of fossil fuels by industry and consumers).
24 Union of Concerned Scientists, 2015. The Climate Deception Dossiers (2015). Retrieved from: www.ucsusa.org/global-warming/fight-misinformation/climate-deception-dossiers-fossil-fuel-industry-memos#.XGyHe-j7Q2x [Accessed 20 August 2018].
25 Berners-Lee, M. and Clark, D., 2013. *The Burning Question.* London: Profile Books.
26 Huber, M., 2013. *Lifeblood: Oil, Freedom and the Forces of Capital.* Minneapolis: University of Minnesota Press.
27 Di Muzio, T., 2015. *Carbon Capitalism: Energy, Social Reproduction and World Order.* London: Rowman & Littlefield.
28 ibid.
29 Church, R., 1986. *The History of the British Coal Industry: Victorian Pre-Eminence. Volume 3.* Oxford: Oxford University Press; Supple, B., 1987. *The History of the British Coal Industry: The Political Economy of Decline. Volume 4.* Oxford: Clarendon Press, 361–424; Goodchild, J., 1978. *The Coal Kings of Yorkshire.* Wakefield: Wakefield Historical Publications.
30 McCord, N., 1979. *North East England: An Economic and Social History.* London: Batsford Academic.
31 Phillips, K., 2003. *Wealth and Democracy: A Political History of the American Rich.* New York: Broadway Books.
32 World Resources Institute. CAIT Climate Data Explorer. Retrieved from: www.wri.org/our-work/project/cait-climate-data-explorer [Accessed 20 August 2018].
33 Deloitte Access Economics, 2015. Consumption-based Carbon Emissions. Retrieved from: www2.deloitte.com/content/dam/Deloitte/au/Documents/Economics/deloitte-au-economics-carbon-analytics-consumption-based-carbon-emissions-050815.pdf [Accessed 20 August 2018].

34 Department for Business, Energy and Industrial Strategy, 2018. Final UK Greenhouse Gas Emissions National Statistics: 1990–2016. *National Statistics*. Retrieved from: www.gov.uk/government/statistics/final-uk-greenhouse-gas-emissions-national-statistics-1990-2016[Accessed 20 December 2018].
35 Committee on Climate Change, 2017. Quantifying Greenhouse Gas Emissions. *CCC*. Retrieved from: www.theccc.org.uk/wp-content/uploads/2017/04/Quantifying-Greenhouse-Gas-Emissions-Committee-on-Climate-Change-April-2017.pdf [Accessed 20 December 2018].
36 Chakravarty, S. et al., 2009. Sharing Global CO2 Emission Reductions among One Billion High Emitters. *PNAS*, [online], 106 (29). Retrieved from: doi:10.1073/pnas.0905232106 [Accessed 17 February 2018].
37 Fischer, D., 2009, 6 July. Who's to Blame? Making Poor Nations Share the Cost of Fighting Climate Change. *Scientific American*. Retrieved from: www.scientificamerican.com/article/poor-pay-for-climate-change/ [Accessed 20 November 2018].
38 Baer et al., 2009. Greenhouse Development Rights Framework. *EcoEquity*. Retrieved from: http://gdrights.org/wp-content/uploads/2009/01/thegdrsframework.pdf [Accessed 20 December 2018].
39 CSO Equity Review, 2018. After Paris: Inequality, Fair Shares, and the Climate Emergency. Retrieved from: doi:10.6084/m9.figshare.7637669 [Accessed 20 December 2018].
40 Crutzen, P. J. and Stoermer, E. F., 2000. The "Anthropocene". *The International Geosphere–Biosphere Programme (IGBP) Newsletter*, [online], 41 (May 2000). Retrieved from: www.igbp.net/download/18.316f18321323470177580001401/1376383088452/NL41.pdf [Accessed 16 August 2018].
41 Steffen, W. et al., 2011. The Anthropocene: Conceptual and Historical Perspectives. *Philosophical Transactions of the Royal Society*, [online], 369 (1938). Retrieved from: doi:10.1098/rsta.2010.0327 [Accessed 17 February 2018].
42 Bonneuil, C. and Fressoz, J-B., 2017. *The Shock of the Anthropocene: The Earth, History and Us*. London: Verso.
43 Moore, J. W., 2017. The Capitalocene, Part I: On the Nature and Origins of Our Ecological Crisis. *The Journal of Peasant Studies*, [online], 44 (3). Retrieved from: doi:10.1080/03066150.2016.1235036 [Accessed 17 February 2018].
44 Gaffney, O. and Steffen, W., 2017. The Anthropocene Equation. *The Anthropocene Review*, [online], 4 (1). Retrieved from: doi:10.1177%2F2053019616688022 [Accessed 17 February 2018].
45 Malm, A., 2016. The Anthropocene Myth. *Jacobin Magazine*. Retrieved from: www.jacobinmag.com/2015/03/anthropocene-capitalism-climate-change/ [Accessed 20 November 2018].
46 Moore, J. W., 2015. The Rise of Cheap Nature. *In*: Moore, J. W, ed. *Anthropocene or Capitalocene?: Nature, History, and the Crisis of Capitalism*. Oakland, CA: PM Press, 169.
47 Ulvila, M. and Wilén, K. B., 2018. Engaging with the Plutocene. *Radical Ecological Democracy*. Retrieved from: www.radicalecologicaldemocracy.org/engaging-with-the-plutocene/ [Accessed 20 December 2018].
48 Griffin, P., 2017. The Carbon Majors Database CDP Carbon Majors Report 2017. *Carbon Disclosure Project*. Retrieved from: https://b8f65cb373b1b-7b15feb-c70d8ead6ced550b4d987d7c03fcdd1d.ssl.cf3.rackcdn.com/cms/reports/documents/000/002/327/original/Carbon-Majors-Report-2017.pdf [Accessed 20 August 2018].

3 The polluter elite and moral responsibility

Even though the *polluter elite* are more responsible for climate change, they are less likely to directly feel the impact of extreme weather events. This chapter looks at why the richest are able to better avoid the consequences of their pollution. It also discusses the moral questions around profiting from an economic activity that causes pollution by questioning whether the *polluter elite* are really wealth creators.

3.1 Environmental injustice

The *polluter elite* are less likely to suffer the consequences of their pollution in the short or long-term. Even if some of the richest people have not been able to fully escape the impact of climate change through hurricanes, floods or heat waves, they can recover more quickly. A number of studies have established the links between climate change impacts and inequality, in particular with regard to heightened vulnerability. There is a vicious cycle at play within both developed and developing countries. Those groups who are already poorer tend to disproportionately suffer from negative impacts of climate change due to their heightened exposure and unequal ability to cope.[1] This then deepens their deprivation. Boyce argues that when there are extreme weather events linked to climate change it is the poorest who are most vulnerable because they are already living hand to mouth, for example, lacking access to health care.[2] There is a global environmental injustice in that the countries of the global south which have contributed the fewest emissions are the most vulnerable and affected by extreme weather events.[3]

The IPCC's 2018 report notes that "in all societies" it is important to recognise the negative impact of global warming on the poor and disadvantaged and therefore the "consideration of ethics and equity can help address the uneven distribution of adverse impacts". It is to the uneven impacts of climate change in the US and the UK that we now turn.

A striking example of this dynamic was the aftermath of Hurricane Katrina which hit New Orleans in 2005, when people of black and other minority ethnic backgrounds were disproportionately impacted. A variety of economic and political factors influenced the fact that low-income African American communities were living in the low-lying areas of the city.[4] Prior research had showed that compared to white people, African American communities often lived in low-lying topographic areas in the 146 cities in the US, and that the ethnic communities were more vulnerable to hurricanes, floods and earthquakes. The hurricane illustrated how existing inequalities along economic and racial lines led to a reduction in resources for recovery being allocated to areas where disadvantaged groups live, which further deepened those inequalities.[5] Disadvantaged groups received lesser public resources than what they needed to cope and recover. Although areas in New Orleans inhabited by low income and the black population suffered worse damage, the public recovery efforts in these areas proceeded at much slower rates than in areas inhabited by the wealthier and white population. In the rare cases where a well-off neighbourhood was in a low-lying area, such as the Lakeview neighbourhood which had one of the lowest elevations in New Orleans Parish, it was still able to recover much more quickly than other areas, due to, in part, the relative wealth of that community. In summary, Hurricane Katrina symbolised the interlacing between social difference and environmental risk.

In the UK flooding linked to climate change is increasing in frequency. In England there is a correlation between areas at higher risk of flooding, such as the coast and riverside estuaries, with concentrations of more deprived communities.[6] These findings are broadly confirmed in other studies that have found that poorer groups are more vulnerable to flooding than the middle classes in all parts of England and Wales, except for the Midlands. Fielding has looked in further detail at which groups are more vulnerable and found that it is non-white communities in deprived regions, and in particular in Wales, who suffer the most from exposure to flood risk.[7] Thus, far from climate change being something that we all experience together, the richest have an unequal ability to adapt.

This is also the case within the US where research since the 1980s has shown a strong correlation between where non-white low-income groups live and exposure to industrial air pollution.[8] This situation was a factor in the emergence of the environmental justice movement and also in the creation of the Office of Environmental Justice by the Environmental Protection Agency (EPA) in 1992 to focus on the environmental and health challenges that minority, low-income, tribal and indigenous peoples face. The current Trump administration has actively weakened the ability of the EPA to enforce environmental legislation and policy, including environmental justice. This is what the Bush administration (2000–2008) did as well from 2001 when it reduced funding.

In the UK research in 1999 by Friends of the Earth showed that poorer areas suffered more from closer proximity to industrial pollution.[9] This was subsequently confirmed in an Environment Agency report in 2002. The concept of environmental justice spread from the US to the UK rising to prominence in the early 2000s following years of campaigning and research by pressure groups. Several studies since have found a correlation between higher levels of air pollution and poorer neighbourhoods. One study carried out for the government found that concentrations of fine particulates and nitrogen dioxide were higher in deprived areas of Greater London, Birmingham City and Greater Belfast.[10] Nitrogen dioxide pollution is a contributor to 23,500 early deaths a year[11] and in London has reduced the capacity of children's lungs by around 5% when it has been above legal levels.[12] Another study of selected neighbourhoods in England found a correlation between higher concentrations of particulate matter (PM10) and nitrogen dioxide (NO2) in the most deprived 20% of neighbourhoods, which were often non-white areas.[13] In summary, the UK's most deprived communities are more likely to live in areas of poor environmental equality, exposed to more risks and therefore have differential access to environmental goods. While the focus is often on deprived communities, it is important to note that there is a large overlap between low-income groups and black and minority ethnic groups.

3.2 Environmental injustice: why do the richest suffer less?

Boyce argues that the richest have a personal interest in protecting the environment but that this is weakened by several factors.[14] They realise that it is still possible for them to live in a healthy environment if they transfer their environmental costs onto others, whether that is within their countries such as the US and UK, or overall to the global south. As Chapter 4 highlights, high economic inequality reinforces political inequality. When there are high inequalities between groups the wealthiest (the winners) are able to shift environmental costs onto weaker actors in society (the losers), including future generations, because they are more powerful.[15] The rich are able to continue benefiting from environmental degradation (for example, via their *investment emissions*). The result is that the pollution of ecosystems and the consequences of climate change are shifted onto the majority of the global population, particularly hitting people living in the global south.

The greater the economic inequality the more likely this situation is to persist, with the poorest suffering more from pollution[16] as the rich carry on shifting their negative environmental impact on to them and using their greater personal resources to protect themselves when they are affected.[17] The examples above on how the poor were more affected by Hurricane Katrina in the US and flooding in the UK, as well as the overall air pollution, highlight that one of the key factors is to what extent someone can avoid the full consequences of pollution by living elsewhere. Under the current economic system in the US and the UK you get what you can pay for (epitomised by the concept of willingness to pay). As wealth further concentrates in the hands of the richest, they increase their ability to live in clean environments and to avoid the full impact of air pollution and extreme weather events. There are some things they cannot avoid entirely, such as floods and wildfires,[18] but they can afford to escape and are more likely to have insurance. In summary, environmental injustices are part of a long-standing process of conflict and negotiation whereby people who control scarcer resources (such as wealth and power) are able to deprive others of access.[19]

As well as avoiding pollution and its effects, it is also the richest who profit the most from activity that harms the environment. If governments regulated to protect the environment (for example, by putting restrictions on pollution in order to have cleaner air and water), the companies the richest invest in would be less profitable because they would have to install expensive equipment and/or pay fines. This leaves them with the dilemma of protecting the environment and their objective of higher returns. The result is that at times of higher economic and political inequality, when profits are very high, the richest people push for a decision on environmental protection to be delayed.[20] Research shows that in the US the richest participate more in politics and are less supportive of protecting the environment.[21] As Chapter 4 documents, some sections of the *polluter elite* actively seek political influence to block environmental policies to maintain the high pollution status quo. When there is higher inequality the richest are more able to use their political influence to protect their economic interests that harm the environment.[22] Building on Boyce's work, Downey concludes that it is undemocratic decision-making processes (at the national and international level) that enable a rich minority to achieve socially and environmentally harmful goals, including shifting the costs on benefit financially from the exploitation of public goods and environmental resources.[23]

3.3 Disconnection can be fatal

Due to their extreme wealth the richest people may be able to avoid experiencing the consequences of climate change and the sixth mass extinction in their daily lives. This could mean they do not see the urgency of changing their consumption habits and investments in polluting companies.

Extreme inequality is leading to a situation where the richest are becoming more and more disconnected from the rest of society as they live in a bubble where they mainly come into contact with other wealthy people. Increasingly, the very richest live in exclusive residences (often gated) and some even own their own islands. They use private transport (cars, private jets, yachts, even submarines). They pay to use private healthcare, eat at exclusive restaurants and attend exclusive events such as the Davos World Economic Forum and the Singapore Yacht Show. The point is not that rich individuals have no idea about environmental issues; it is that they are less likely to face this reality in their day jobs which may contribute to them being more disconnected.

When previous civilizations collapsed one common driver has been that the elite were able to insulate themselves from the impact of their decisions. Often the elite were motivated to seek personal profit even if in doing so they harmed the rest of society.[24] Building on this research Mackay argues that even when societies have possessed sufficient technology and cultural knowledge, they have not used these solutions because the oligarchy has blocked them. Instead, the oligarchy has captured decision-making to enrich themselves and strengthen their own power.[25] Some scholars have suggested educating the super-rich so that they understand that the multiple crises linked to climate change such as water scarcity, climate refugees and conflict will one day affect them.[26] But educating the *polluter elite* (climate change and inequality have been on the agenda at the Davos World Economic Forum in the last few years) is unlikely to work when they profit from pollution.

3.4 Is it morally questionable to profit from pollution?

The richest people in the US and the UK (and in many other countries) hold shares in highly polluting companies because it is profitable. In 2017 oil and gas companies were among the top 100 global companies including several of those mentioned in the *polluter elite* database such as ExxonMobil (7th), Shell (23rd), Chevron (27th), Total (52nd) and BP (61st).[27] When compared to renewable energy, fossil fuels are still currently more profitable. The consultancy firm, Wood Mackenzie, found that traditional oil and gas projects give a return of 20% compared to solar or wind projects which were more in the range of 5–9%.[28] Another indicator of the regular profitability of fossil fuels is that so many pension funds are invested in them precisely because they often pay reliable dividends.

As Jacobs and Mazzucato note: "throughout capitalism's history economic growth has been accompanied by environmental damage, from the pollution of air, water and land to loss of habitats and species, a constant subtraction from its successes in increasing welfare" (Jacobs and Mazzucato, 2016).[29] In a capitalist economic system, investors seek returns on their capital. If shares in fossil fuels are profitable they will continue to be attractive to investors. As long as it is profitable to exploit new sources, even if they are more difficult and expensive, such as tar sands and drilling in the Arctic, they will continue to receive investment.

These are not necessarily moral decisions. The executives who run these companies probably also have children and grandchildren whom they want to lead healthy lives. It is unlikely that the individuals within the *polluter elite* choose to specifically profit from greenhouse gas emissions (even if there are people who are seeking to profit from the consequences of global warming by selling goods and services to adapt). They just do not see it as a problem that their source of wealth rests on the operations of the companies that pollute. If they did then they would not reach a senior position in the company. What investors in oil, gas and coal companies care about are the company's proven reserves as an indicator of its ability to prove it will have higher earnings in the future than its competitors. The priority becomes profit ahead of human health. For example, the factory owners in Manchester continued to burn coal despite it causing air pollution in the 1840s that led to people dying from respiratory diseases.[30]

As fossil fuels are profitable, it means that if the individuals listed in the polluter database divested by selling their shares, they would be replaced by other individuals who will take advantage of this opportunity (a constant critique of the fossil fuel divestment movement). Investments will go where there are profits to be made. This could be in fossil fuels and it could also be renewable energies.

The profitability of fossil fuels means that although the price of oil, gas and coal can fluctuate wildly, they are still seen as a safe bet and for this reason are often part of a diversified portfolio. Therefore, when the *polluter elite* invest in fossil fuels it is partly to spread the risk of their portfolio with the ultimate aim of protecting their existing wealth and hoping it will grow larger at a fast enough rate to keep them wealthy compared to their peers.

A motivating factor for holding such reliably profitable investments is that the *polluter elite* can use the income from their shareholdings to fund their luxury high-carbon lifestyle (see Chapter 1). The more money they have the more they can compete in status consumption. Luxury lifestyle status competition is one expression of wealth status competition. As inequality of wealth and income increases the richest have ever more money to fund this lifestyle.

3.4.1 The victims of climate change

Chapter 2 argued that the *investment emissions* of the *polluter elite* mean they hold greater historical responsibility for global warming. These investments are in oil, gas and coal multinationals who operate internationally (whether they are headquartered in the US, the UK or countries in Asia) and therefore it is appropriate to discuss their global responsibility. The distinction between direct and indirect shareholdings was discussed in Chapter 2 in terms of levels of agency and responsibility and complicity. But what matters is that both types of shareholdings profit from company operations that generate pollution. This could be considered to be morally questionable.[31] This is because pollution leads to more deaths both from immediate effects, such as respiratory diseases linked to air pollution, and from longer-term effects such as extreme weather events.[32] The science on climate change has become more and more established since 1990 when the IPCC released its first report. That means that over the last 30 years when the *polluter elite* have invested in new fossil fuel extraction and infrastructure such as oil wells, power plants, roads, airports etc they have known it will contribute to rising emissions.

The IPCC's 2018 report highlights that globally certain groups such as indigenous people and local communities who depend on agricultural or fishing for their livelihood are at "disproportionately higher risk of adverse consequences of global warming". This is particularly the case for Arctic ecosystems, dryland regions, small islands and the poorest countries. These are cases of global environmental injustice that are happening alongside environmental injustice within countries as was documented above for the US and the UK. In effect, the *unequal ability to pollute* through a high-carbon lifestyle and *investment emissions* results in the richest people spreading their pollution in the US and the UK, and globally. The richest people tend to pollute "somewhere else" and are very unlikely to have to bear the cost of that pollution. This is perhaps another reason to look at the *polluter elite* at the global level in addition to closer scrutiny within countries such as the US and the UK.

There is a growing understanding of how climate change will lead to a greater number of deaths in the future. Between 1993 and 2012 around 530,000 died due to the impact of approximately 15,000 extreme weather events.[33] Whilst it is difficult to say these were all linked to climate change what is more certain is that global warming is increasing the frequency and intensity of these events such as hurricanes, tropical cyclones, flooding, droughts and heat waves. The World Health Organisation predicts that between 2030 and 2050 an extra 250,000 people will die each year from factors such as heat stress, malnutrition, malaria and diarrhoea.[34] One of the findings of the IPCC's 2018 report is that there will be more heatwaves in cities and higher risks of malaria and dengue fever. Studies on the impact of climate change on health predict increasing heat waves will lead to higher mortality rates, particularly in tropical and subtropical regions.[35] The Lancet Commission on Planetary Health study also found that under high emission scenarios increasing global temperatures would lead to higher death rates, but that this would depend on the region in question. For example, northern Europe, East Asia, and Australia would be less affected compared to warmer regions, such as the central and southern parts of America or Europe, and especially southeast Asia.[36]

The 2017 Lancet Commission on Pollution and Health concludes that pollution from power plants (in particular burning of coal), chemical production, mining and deforestation is one of the "main anthropogenic drivers of global climate change".[37] This pollution which is a threat to the health of the planet is directly linked to climate change and "disproportionately kills the poor and the vulnerable" as just over 90% of pollution-related deaths happen in middle and lower-income countries. Based on data from the World Health Organisation and other sources the Commission concludes that diseases caused by pollution led to around nine million dying prematurely in 2015 (16% of global deaths). The main causes were industrial emissions, vehicular exhaust, and toxic chemicals which produced air and water pollution. Deaths from indoor pollution are due to the burning of wood and other similar materials. According to World Health Organisation data, outdoor air pollution is a cause of over four million early deaths a year. The majority of these deaths are in countries such as India and China.[38]

Reflecting on new gas finds in the North Sea in 2019, the head of Friends of the Earth, Craig Bennet, said:

> We need to be really clear about this – if these companies were to exploit this new find, and if these fossil fuels were to get burned, that will mean many more people will die because of the impacts of climate change.[39]

The impact of pollution, like climate change, is a gradual process meaning its fatal impact is not always immediately visible. It could be described as slow violence because it is incremental, dispersed across time and space, and often not perceived as a type of violence.[40] For example, oil spillages and gas flaring causing acid rain have contributed to making the Niger Delta, the breadbasket of the country, less fertile.

The *unequal ability to pollute* and the fact that pollution can be fatal highlight how climate change impacts on the human right to life. The United Nations Office of the High Commissioner for Human Rights highlighted in a 2007 report that "climate change poses a direct threat to a wide range of universally recognised, fundamental rights, such as the rights to life, food, adequate housing, health and water". Caney advocates for a human-rights approach which acknowledges this and would therefore "condemn as unjust a situation in which some (who are advantaged) expose others (who are vulnerable) to risks that threaten the latter's basic interests".[41]

3.5 From wealth creators to wealth destroyers?

I think it is time to question if polluting activities are a form of value creation. I argue that when the richest people profit from pollution this should not be seen as a legitimate way to make money. Yes, there is wealth creation in that a profit is made from economic activities such as oil, gas and coal and this is returned to shareholders as dividends. These companies also create direct and indirect jobs. But this wealth creation is based on pollution that is contributing to global warming and can have fatal consequences through air pollution and extreme weather events. As part of a shift in societal values to see profit from pollution as immoral this should include a shift from seeing the *polluter elite* as wealth creators to wealth destroyers, where wealth is understood as the necessary conditions for a habitable planet. In recognition of the fact that the *polluter elite* make money from pollution, I propose an additional column in the global rich lists which clearly shows this. This would distinguish them as being decision makers (responsibility for a portion of the company's direct emissions) or just shareholders (complicity in the company's direct emissions).

Table 3.1 Richest Billionaires in the US

Ranking	*Name*	*Net worth*	*Source*
1	Jeff Bezos	$112 billion	Amazon
2	Bill Gates	$90 billion	Microsoft
3	Warren Buffet	$84 billion	Berkshire Hathaway
8	David Koch	$60 billion	Koch Industries
8	Charles Koch	$60 billion	Koch Industries

Source: Forbes.[42]

The current rich list in the US is shown in Table 3.1.

Table 3.2 Sample from Proposed Rich List

Global ranking	*Name*	*Net worth*	*Source*	*Pollution*
1	Jeff Bezos	$112 billion	Amazon	Unknown.
2	Bill Gates	$90 billion	Microsoft	Unknown.
3	Warren Buffet	$84 billion	Berkshire Hathaway	Owner of NetJets airline company. As a decision maker he is responsible for a portion of the company's direct emissions. Investments in companies facilitating fossil fuels. Owner of BNSF freight trains which transport oil.
8	David Koch	$60 billion	Koch Industries	Co-owner of Koch Industries Estimated annual emissions of 24 million metric tons CO2e in 2011. As a decision maker he is responsible for a portion of the company's direct emissions.

Global ranking	*Name*	*Net worth*	*Source*	*Pollution*
8	Charles Koch	$60 billion	Koch Industries	Co-owner of Koch Industries Estimated annual emissions of 24 million metric tons CO2e in 2011. As a decision maker he is responsible for a portion of the company's direct emissions.

This is what I propose a polluter rich list should look like (Table 3.2):

Table 3.3 Sample from Proposed Polluter Rich List

Global ranking	*Name*	*Fortune*	*Source*	*Pollution*
73	Carl Icahn	$17.7 billion	Investments	On 31 December 2015 he held 73,050,000 shares in Chesapeake Energy Corporation (prominent fracking company) meaning he was *complicit* in 504,428 metric tons of CO2e.

Source: Forbes billionaire list;[43] *Investment emissions* calculation from the *polluter elite* database.

And when a specific shareholding is known there would be the percentage of company emissions. For example (Table 3.3).

This idea could be extended beyond climate change to look at which individuals have profited from species extinction. For example, a study published in early 2019 found that 40% of insect species are declining and threatened with extinction.[44] According to the study, the main drivers of this decline are intensive agriculture (including pesticides), urbanisation and climate change. In the case of intensive agriculture, it would be possible to identify agribusiness companies using pesticides in large quantities. The next step would be to identify who ran the company and who profited from this economic activity through a salary and shareholding. The decision makers who run the company hold the most responsibility. Other shareholders are complicit.

3.6 Conclusion

This chapter has documented the environmental injustice in the US and the UK. The poorest are hit hardest by air pollution and extreme weather events such as Hurricane Katrina and flooding in the UK. Meanwhile, the richest are disconnected from the harsh realities of climate change and have more capacity to insulate themselves from the consequences. In the age of global warming and species extinction is it still accurate to refer to rich individuals who run polluting companies as wealth creators?

Notes

1 Cutter, S. et al., 2003. Social Vulnerability to Environmental Hazards. *Social Science Quarterly*, [online], 84 (2). Retrieved from: doi:10.1111/1540-6237.8402002 [Accessed 17 February 2018].
2 Boyce, J. K., 2013. *Economics, the Environment and our Common Wealth*. Northampton, MA: Elgar, 15.
3 People's Demands for Climate Justice. Homepage. Retrieved from: www.peoplesdemands.org/?fbclid=IwAR3ZJ7Q6Sv7FD1ro9uXZkYVVP-JrO3-e6fsVE6CdgjVvh7NL3Zf0cDP3WjVo [Accessed 9 March 2019].
4 Mutter, J., 2015. *Disaster Profiteers: How Natural Disasters Make the Rich Richer and the Poor Even Poorer*. Bristol: Bristol University Press.
5 Islam, S. N. and Winkel, J., 2017. Climate Change and Social Inequality. Department of Economic & Social Affairs Working Paper. *UN*. Retrieved from: www.un.org/esa/desa/papers/2017/wp152_2017.pdf [Accessed 20 December 2018].
6 Walker, G. et al., 2006. Addressing Environmental Inequalities: Flood Risk. *Environment Agency*. Retrieved from: www.staffs.ac.uk/assets/SC0 20061_SR1%20report%20-%20inequalities%20flood%20risk_tcm44-21951.pdf [Accessed 20 December 2018].
7 Fielding, J. L., 2017. Flood Risk and Inequalities between Ethnic Groups in the Floodplains of England and Wales. *Disasters*, [online], 42 (1). Retrieved from: doi:10.1111/disa.12230 [Accessed 17 February 2018].
8 Boyce, K. et al., 2014. Three Measures of Environmental Inequality. INET Working Paper. Retrieved from: www.ineteconomics.org/uploads/papers/WP12-Boyce-et-al.pdf [Accessed 20 December 2018].
9 Dyer, R., 2015. The Environmental Reasons for Reducing Inequalities. *Friends of the Earth*. Retrieved from: https://friendsoftheearth.uk/sites/default/files/downloads/environmental-reasons-reducing-inequalities-90784.pdf [Accessed 20 November 2018].
10 Pedersen, O. W., 2011. Environmental Justice in the UK: Uncertainty, Ambiguity and the Law. *The Lancet*, [online], 391 (10119). Retrieved from: doi:10.1016/S0140-6736(17)32345-0 [Accessed 17 February 2018].

11 Carrington, D., 2018, 5 November. Air Pollution: Everything You Should Know about a Public Health Emergency. *The Guardian*. Retrieved from: www.theguardian.com/environment/2018/nov/05/air-pollution-everything-you-should-know-about-a-public-health-emergency [Accessed 21 December 2018].
12 Mudway, I. et al., 2018. Impact of London's Low Emission Zone on Air Quality and Children's Respiratory Health: A Sequential Annual Cross-sectional Study. *The Lancet: Public Health*, [online], 4 (1). Retrieved from: doi:10.1016/S2468-2667(18)30202-0 [Accessed 17 December 2018].
13 Fecht, D. et al., 2014. Associations between Air Pollution and Socioeconomic Characteristics, Ethnicity and Age Profile of Neighbourhoods in England and the Netherlands. *Environmental Pollution*, [online], 198 (201–210). Retrieved from: doi:10.1016/j.envpol.2014.12.014 [Accessed 17 February 2018].
14 Boyce, J. K., 1994. Inequality as a Cause of Environmental Degradation. *Ecological Economics*, [online], 11 (3). Retrieved from: doi:10.1016/0921-8009(94)90198-8 [Accessed 17 February 2018].
15 Boyce, J. K., 2018. How Economic Inequality Harms the Environment. *Scientific American*. Retrieved from: www.scientificamerican.com/article/how-economic-inequality-harms-the-environment/ [Accessed 20 December 2018].
16 Knight, K. et al., 2017. Wealth Inequality and Carbon Emissions in High-income Countries. *Social Currents*, [online], 4 (5). Retrieved from: doi:10.1177%2F2329496517704872 [Accessed 17 February 2018].
17 Laurent, E., 2014. Inequality as Pollution, Pollution as Inequality: The Social-ecological Nexus. *Stanford Centre on Poverty and Inequality* [online]. Retrieved from: http://web.stanford.edu/group/scspi/_media/working_papers/laurent_inequality-pollution.pdf [Accessed 6 October 2018].
18 Chokshi, N., 2018, 13 November. Neil Young and Miley Cyrus among Celebrities Who Lost Homes in California Wildfires. *The New York Times*. Retrieved from: www.nytimes.com/2018/11/13/us/celebrities-lost-homes-california-fires.html [Accessed 21 December 2018].
19 Walker, G., 2012. *Environmental Justice: Concept, Evidence and Politics*. London: Routledge.
20 Boyce, J. K., 2002. *The Political Economy of the Environment*. Northampton, MA: Elgar.
21 Knight, K. et al., 2017. Wealth Inequality and Carbon Emissions in High-income Countries. *Social Currents*, [online], 4 (5). Retrieved from: doi:10.1177%2F2329496517704872 [Accessed 17 February 2018].
22 Cushing, L. et al., 2015. The Haves, the Have-Nots, and the Health of Everyone: The Relationship between Social Inequality and Environmental Quality. *Annual Review of Public Health*, [online], 36 (193–209). Retrieved from: www.annualreviews.org/doi/10.1146/annurev-publhealth-031914-122646 [Accessed 17 February 2018].
23 Downey, L. A., 2017. *Inequality, Democracy, and the Environment*. New York: New York University Press.
24 Diamond, J. 2011. *Collapse: How Societies Choose to Succeed or Fail*. New York: Penguin.
25 Mackay, K., 2017. *Radical Transformation: Oligarchy, Collapse, and the Crisis of Civilization*. Toronto: Between the Lines Books.

26 Otto, L. et al., 2019. Shift the Focus from the Super-poor to the Super-rich. *Nature Climate Change*, [online], 9 (82–84). Retrieved from: www.nature.com/articles/s41558-019-0402-3 [Accessed 17 February 2019].
27 PwC, 2017. Global Top 100 Companies by Market Capitalisation. Retrieved from: www.pwc.com/gx/en/audit-services/assets/pdf/global-top-100-companies-2017-final.pdf [Accessed 20 November 2018].
28 Financial Times. 2018, 13 November. Oil Majors Switch on to a Future in Power Generation. *Financial Times.* Retrieved from: www.ft.com/content/699584f4-e36e-11e8-a6e5-792428919cee [Accessed 21 December 2018].
29 Jacobs, M. and Mazzucato, M., 2016. Rethinking Capitalism: An Introduction. *In*: Jacobs, M. and Mazzucato, M., eds. *Rethinking Capitalism: Economics and Policy for Sustainable and Inclusive Growth.* Chichester: Wiley-Blackwell, 2.47.
30 Malm, A. 2016. *Fossil Capital: The Rise of Steam Power and the Roots of Global Warming.* London: Verso.
31 Go Fossil Free. Not a Penny More for Fossil Fuels. Retrieved from: https://gofossilfree.org/not-a-penny-more/ [Accessed 20 August 2018]. The fossil fuel divestment movement follows this logic:

> the more we can make climate change a deeply moral issue, the more we will push society towards action. We need to make it clear that if it's wrong to wreck the planet, then it's also wrong to profit from that wreckage.

32 Aronoff, K., 2019. It's Time to Try Fossil-Fuel Executives for Crimes against Humanity. *Jacobin Magazine.* Retrieved from: www.jacobinmag.com/2015/03/anthropocene-capitalism-climate-change/ [Accessed 10 March 2019].
33 Hashim, J. and Hashim, Z., 2015. Climate Change, Extreme Weather Events, and Human Health Implications in the Asia Pacific Region. *Asia Pacific Journal of Public Health*, [online], 28 (2). Retrieved from: doi:10.1177%2F1010539515599030 [Accessed 17 February 2019].
34 World Health Organization. Climate Change and Health. Retrieved from: www.who.int/news-room/fact-sheets/detail/climate-change-and-health [Accessed 20 August 2018].
35 Guo, Y. et al., 2018. Quantifying Excess Deaths Related to Heatwaves Under Climate Change Scenarios: A Multicountry Time Series Modelling Study. *Plos Medicine*, [online], 15 (7). Retrieved from: doi:10.1371/journal.pmed.1002629 [Accessed 17 February 2019].
36 Whitmee, S. et al., 2015. Safeguarding Human Health in the Anthropocene Epoch: Report of the Rockefeller Foundation–Lancet Commission on Planetary Health. *The Lancet*, [online], 386 (10007). Retrieved from: doi:10.1016/S0140-6736(15)60901-1 [Accessed 17 February 2018].
37 Landrigan, P. J. et al., 2017. The Lancet Commission on Pollution and Health. *The Lancet*, [online], 391 (10119). Retrieved from: doi:10.1016/S0140-6736(17)32345-0 [Accessed 17 February 2018].
38 Carrington, D., 2018, 5 November. Air Pollution: Everything You Should Know about a Public Health Emergency. *The Guardian.* Retrieved from: www.theguardian.com/environment/2018/nov/05/air-pollution-everything-you-should-know-about-a-public-health-emergency [Accessed 21 December 2018].

39 Gabbatiss, J. and Gregory, A., 2019, 30 January. Green Campaigners Condemn 'Disgraceful' North Sea Gas Discovery Hailed as Biggest in a Decade. *The Independent*. Retrieved from: www.independent.co.uk/environment/north-sea-gas-sea-oil-fossil-fuels-drilling-climate-change-global-warming-a8754156.html [Accessed 17 February 2019].
40 Nixon, R. 2011. *Slow Violence and the Environmentalism of the Poor*. Cambridge, MA: Harvard University Press.
41 Caney, S., 2010. Climate Change, Human Rights and Moral Thresholds. *In*: Gardiner, S. et al., eds. *Climate Ethics*. Oxford: Oxford University Press, 170.
42 Forbes. Billionaires List. Retrieved from: www.forbes.com/billionaires/list/#version:static [Accessed 9 March 2019].
43 Forbes. Billionaire Profile: Carl Icahn. Retrieved from: www.forbes.com/profile/carl-icahn/#6502be3d35eb [Accessed 9 March 2019].
44 Sánchez-Bayo, F. and Wyckhuys, A. G., 2018. Worldwide Decline of the Entomofauna: A Review of Its Drivers. *Biological Conservation*, [online], 27 (6). Retrieved from: doi:10.1016/j.biocon.2019.01.020 [Accessed 17 February 2019].

4 The decisive political influence of the polluter elite

This chapter looks at how and why the *polluter elite* obtain political influence. It seeks to understand why the political power of the *polluter elite* has been so effective by exploring their relationship with the state. The *polluter elite* have had significant influence on whether the low-carbon transition has accelerated or slowed since 1990. This is particularly evident in the US, and to a lesser extent in the UK.

4.1 The increasing focus on the actors blocking sustainability transitions

Over the past few decades, there has been growing consensus among climate scientists on the urgency to reduce greenhouse gas emissions. In this context a rapidly expanding field of research has appeared which looks at sustainability transitions, governance and innovation and how they can be advanced.[1] In recognition of the dynamic and open-ended nature of transitions, this interdisciplinary field covers multiple elements and multiple actors and attempts to understand the interactions between them.

A major focus of the sustainability transitions literature has been on how green innovations such as wind turbines and electric vehicles emerge and challenge existing technology. A key reference point has been the Multi-Level Perspective. This is a framework to understand large-scale transitions through the interaction between the actors already in power (referred to as regimes and incumbents), actors who are trying to change things through green innovations (referred to as niches) and exogenous macro-factors such as climate change (this is referred to as the landscape level).[2] The interactions between these different levels, and whether the niche or regime is more influential, shapes how the transition happens. For example, pressures such as climate change could create the opportunity that enables emerging niche-innovations to compete with the existing regime in economic, technical, political and cultural dimensions. Other key frameworks are the Technological Innovation System approach (new technologies are developed, diffused, experimented with, legitimised and financed), Strategic Niche Management (how radical new innovations emerge) and Transition Management (the role of policymakers in shaping change).[3]

The transition from one energy system to another is often portrayed as a protracted process that can take decades to centuries to happen. Although there is potential for a rapid energy transition[4], so far it has not been happening fast enough. The US and UK economies remain heavily dependent on fossil fuels. In the US renewable electricity generation from solar and wind sources grew from 0.1% in 1990 reaching 8% by the third quarter of 2018.[5] In the UK just over 29% of electricity was generated from renewable sources in 2017 (including wind, solar, hydro and biomass).[6] Despite this progress, there is still a lot more to be done. There has been a decline in coal in both countries which has reduced emissions from this source, but this has mainly been replaced by gas rather than renewable energy. Meanwhile, the consumption of petroleum, mainly for transport, has remained at a similar level between 1990 and 2017.

It is in this context that scholars working on sustainability transitions are increasingly recognising that in addition to studying green innovation, there needs to be a greater focus on how the transition could be deliberately accelerated and how powerful incumbents who are blocking change could be addressed.[7] Part of this is deepening understanding of why transitions can fail as incumbents use multiple strategies to adapt and defend their position.[8] The resistance of the incumbent *polluter elite* is no surprise, as vested interests have adopted this strategy in previous energy transitions. In the US the incumbent regime is so well organised that it deserves greater academic scrutiny.[9] In the UK fossil fuel incumbents have resisted the overall decarbonisation of the electricity system meaning that while there has been increasing generation from renewable energy, this has mainly been additional rather than a substitute for fossil fuels. Overall, this situation requires greater attention to how regime resistance undermines progress towards a low-carbon economy, because the expansion of fossil fuels highlighted above is using up the limited carbon budget.[10]

Issues of power and politics are becoming more established in sustainability transitions debates. Why do some actors win or lose?[11] One way to explore this issue is to look at power asymmetries between the powerful regime and the weaker niches.[12] As the section below will show, the *polluter elite* follow the pattern of regime actors holding significant financial and technical resources.[13] Following the logic of the

Multi-Level Perspective this is an example of the existing regime acting to ensure a niche is undermined.[14] For example, the ways in which the Conservative government removed subsidies for solar power are discussed below. As incumbents hold so much power some scholars are exploring how they could be destabilised (this is dealt with in Chapter 5).

4.2 The *polluter elite* seek to defend their net worth

Before detailing how the *polluter elite* seek political influence I will look at why they do this. The *polluter elite* aim to shape state policy because it affects the profitability of their investments and their net worth. They fear the loss of wealth and power. If the state rules against fossil fuels, they will lose their position as wealthy individuals at the top of society, they will lose their ability to seek future political influence, and they will be less able to continue status-seeking high-carbon luxury consumption.

Their shareholding is a motivating factor that helps to explain why they seek political influence (see Figure 4.1). They tend to hold on to these shares for many years. For example, the executive team often

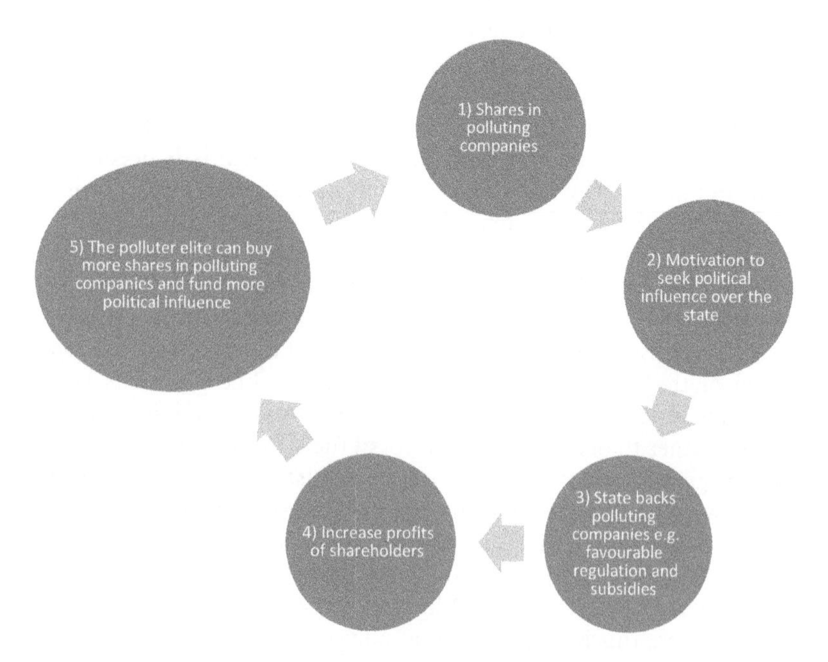

Figure 4.1 Profits from polluting activities as a driver of increasing political influence of the polluter elite.

Source: Author.

spend their entire careers in the fossil fuel industry and directors often hold these positions for around ten years.

The *polluter elite* understand that the state presents a threat and an opportunity. Oil and gas companies depend on the state to uphold property rights and contracts. They must have access to oil and gas to be able to make a profit. They have a clear interest in blocking, or at least slowing down, measures to address climate change as this would hugely affect their operations and profitability. For these reasons those who hold wealth can sometimes end up becoming directly involved in the policy-making process to achieve their goals.[15]

This is because a proven way to defend and enhance their fossil fuel interests is to actively attempt to influence what government does. This is why the companies run by the *polluter elite* decide to influence decisions by donating to political parties and funding lobbyists.[16]

The *polluter elite* aim to influence regulations that could restrict their activity or increase their operating costs. They need to counter the potential threat to their profits by ensuring they can continue extracting. They comprehend that if the government intervenes in the market to reduce emissions then this will result in large-scale economic losses for polluting companies. This is such a grave risk because the huge amount of finance that has already been used to build fossil fuel infrastructure is a sunk investment. In other words, the capital already invested in the rigs, mines, pipelines, railroads, refineries, tankers, etc cannot easily be recovered. Companies will use this infrastructure for as long as they can to maximise their assets. These investors understand there is the risk of stranded assets whereby they could no longer use this infrastructure. It cannot be retrofitted for alternative energy sources such as solar and wind; for example, an oil pipeline cannot be used to transport solar energy.

The first objective for the *polluter elite* is to ensure they still have the green light to continue extracting fossil fuels and maintain all the subsequent stages along the supply chain (refining, distribution, etc) in order to continue to use these sunk investments. Their ability to make a profit is dependent on this green light. One example to illustrate how regulations and policies enacted by the state affect the profitability of these investments and ultimately the net worth of the *polluter elite* is that of Kelcy Warren. As CEO of the pipeline operator Energy Transfer Partners that is building the Dakota Access Pipeline he donated US$100,000 to Trump's Presidential campaign and followed this up with another $250,000 for Trump's inauguration.[17] When Trump arrived in office in early 2017, he approved the Dakota Access Pipeline which was a factor in the company seeing its profits surge from $583 million in 2016 to $2.5 billion in 2017 (Trump sold his previous shares in Energy Transfer Partners before he became President). This decision is likely to have been a factor in Kelcy Warren's net worth jumping from $1.7 billion in March 2016 to $4.5 billion in March 2017. As of February 2019, he was estimated to be worth around $4.5 billion (Figure 4.2).

These wealthy individuals also understand that if government intervention can be turned to their advantage, this represents a business opportunity. Their political power can be used to open up new areas of oil and gas extraction. It is the government that has the ability to make this possible through legislation and other policy measures. For example, the administration of George W. Bush, a millionaire with strong ties to the oil industry, weakened environmental regulations which facilitated greater exploitation of fossil fuels. His government made fracking exempt from the Safe Drinking Water Act (despite scientists at the EPA objecting) which was one driver in the rapid expansion, and thus profitability, of the fracking sector. In the UK the Conservative government, which has been in power

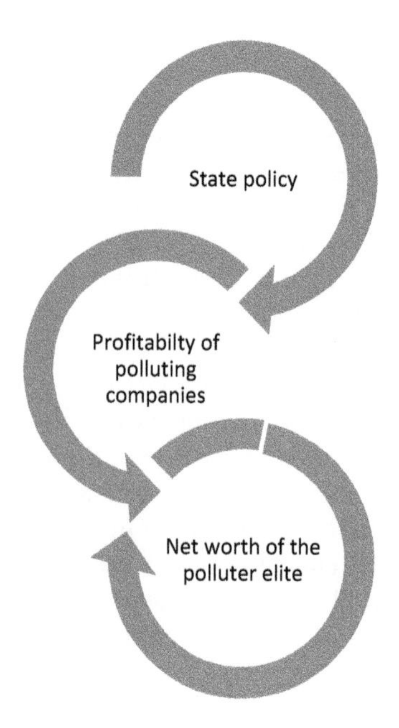

Figure 4.2 The role of the state in influencing the net worth of the polluter elite.

Source: Author.

since 2010, has consistently implemented policies such as relaxation of planning requirements to enable fracking to go ahead. This is despite local government denying planning permission following protests by local communities.[18]

Another goal for the *polluter elite* running large oil and gas multinationals is to weaken potential rivals. The rise of alternative energy sources, such as solar and wind, and more efficient forms of production and consumption could potentially be a direct threat to many types of polluting activity. This broader interest in defending the fossil fuel status quo helps to explain why some polluting companies, and some wealthy individuals (e.g. those coordinated by the Koch brothers, discussed in the next section) fund climate change denial groups. A member of the *polluter elite* benefits from questioning the overall climate science because this delays the shift to alternative energy sources such as solar. This keeps the fossil fuel machine going and keeps economies dependent on this energy source for economic growth. Whilst their ultimate interest is individual, the elite also have a shared interest in blocking action on climate change.

In summary, the *polluter elite* need to use their political influence to capture the state to slow the transition away from fossil fuels because they understand that ultimately this is a question of what their personal net worth will be. In effect what the *polluter elite* are doing is investing their own wealth to grow that wealth still further. This is why the *polluter elite* database records, where possible, the dividend each individual received. Members of the elite make money on their shareholding in polluting companies which they can then use for whatever they wish. This could include donations to political parties. It could also lead to a high-carbon luxury lifestyle.

4.3 How the *polluter elite* seek political influence

As decision makers at polluting companies (owners, executive team and directors) the *polluter elite* are using their political influence in different ways to slow down the transition. Since the 1970s these companies have significantly increased funding to foundations, think-tanks, public relations firms and political donations to influence both governments and public opinion. See Appendix 1 for some indicative examples. This is particularly the case in the US, and to a lesser extent in the UK.

These different ways to obtain political influence can be grouped under the following headings (based on Geels, 2014 and Newell and Paterson, 1998).[19]

- **Access**: Companies have close contact with senior policymakers which leads to government factoring in their interests when formulating policy; for example, the Trump administration has systematically tried to remove or weaken environmental regulations.
- **Set the overall direction**: The goal is for policymakers to accept the interests of companies as their own viewpoint. There might be disagreement about how to get there but there is general agreement about the direction, for example, the narrative in the UK that it makes sense to start new fracking to ensure energy security and create jobs.
- **Direct lobbying on specific issues**: Forms of organised pressure, for example, the Trump administration withdrew from the Paris Agreement.

One of the most emblematic results of lobbying by affluent individuals and their companies are the continued subsidies these companies receive from the US and UK state.[20] The securing of these subsidies, a form of state aid through measures such as tax relief, is a practical expression of the power of the fossil fuel industry.

In the US it is estimated that oil and gas companies received an average annual subsidy of US$4.86 billion between 1918 and 2009. For comparison over the shorter time period of 1994 to 2009 renewable energy received around US$0.37 billion per year.[21] The Energy Bill approved in 2005 under the George W. Bush administration put in place $6 billion in subsidies for oil and gas and another $9 billion for coal companies.[22] Such is the political power of the fossil fuel industry that even when President Obama repeatedly tried to remove some of these subsidies he was unable to do so. In 2015 and 2016 the Federal and State governments provided on average US$20.5 billion in production subsidies. This is estimated to be seven times what was given to renewable energies.[23]

Rich individuals have a strong desire to influence politicians (or to play this role themselves) in order to guarantee their economic interests. When the political elite acts via government institutions it holds a position of what Guttsman called "unrivalled eminence" to decide on outcomes because it can make laws, and crucially also enforce them.[24] Perhaps the most explicit example of this taking place are the actions of President Donald Trump. In terms of access and setting the direction of government policy the *polluter elite* have had repeated success since he was elected President. When a new President is about to take office, these companies donate to their inauguration. Donald Trump received over $10 million from oil, gas and coal companies and members of their executive teams.[25] Since Trump has been President, he has consistently sought to undermine emissions reductions, but he has not always got his way as the courts have ruled against his decisions.

However, between early 2017 and the end of 2018 the Trump administration had successfully eliminated 47 environmental rules mainly related to fossil fuel extraction and emissions, and was trying to eliminate another 31 rules.[26] The elimination of these rules helped to reduce costs for fossil fuel producers. This made them more competitive abroad which is one factor in the rise in US coal exports. Trump approved the Dakota Access and Keystone XL pipelines. He removed regulation on leasing for oil and gas operations on federal lands. He gave the green light for drilling for oil in US coastal waters.

Trump is also a climate change denier and his actions should be seen as the culmination of decades of effort to confuse and delay government policy to address climate change.[27] In November 2018 he rejected his own government agency's Fourth National Climate Assessment which concluded "climate change could cost the US economy hundreds of billions of dollars and thousands of lives each year in future".[28] Singer explains how Trump's approach as climate sceptic focuses on the claim that climate change is a hoax promoted by people such as a "self-serving elite, grant-hungry climate scientists, left-leaning political leaders, and a 'deep state' of government officials" (Singer, 2018).[29] Trump has appointed a number of sceptics to official positions:

- He chose the CEO of ExxonMobil Rex Tillerson, the world's largest oil company which consistently undermined state action on climate change and funded climate scepticism, to run the State Department. He then replaced Tillerson with another climate change denier Mike Pompeo.
- He appointed climate change denier Scott Pruitt to run the EPA. Pruitt resigned in July 2018.

In addition to Trump being influenced by the *polluter elite*, he has also had personal interests in fossil fuels. While it is difficult to obtain an accurate picture of President Trump's current investments past disclosures show he has held shares in a range of polluting companies, including in ten that are listed in the *polluter elite* database. The Executive Branch Personnel Public Financial Disclosure Report he submitted to the Federal Election Commission in July 2015 and July 2016 show that during this time he held shares in ExxonMobil, Anadarko, Shell, Devon Energy, Chevron, Occidental Petroleum, Total, BHP

Billiton, ConocoPhillips and EOG Resources.[30] These forms also show he has held shares in several pipelines including:

- Phillips66 (one project is the Dakota Access pipeline),
- Kinder Morgan's TransMountain pipeline transporting oil from Canada's tar sands,
- TransCanada, developer of the Keystone XL pipeline

BOX 4.1 *POLLUTER ELITE* PROFILE: DAVID AND CHARLES KOCH

David and Charles Koch own Koch Industries, one of the largest private companies in the US. They have each seen their personal wealth rise from an estimated $36 billion in 2013, hitting $60 billion in March 2018, and currently at $51.3 billion in early March 2019.

The Koch brothers have a long track record of funding political parties, fossil fuel lobby groups and climate change denial using their own wealth and via Koch Industries. Based on Centre for Responsive Politics data the DeSmog blog estimates that between 1998 and 2017 Koch Industries spent around $110,520,000 on lobbying for the oil and gas industry.[31] To give a flavour of the size of their political donations, during the 2016 elections the Koch brothers' political network raised $1 billion meaning they had a similar budget to both the Republican and Democratic parties

Citing Environmental EPA figures, Mayer found that in 2011 Koch Industries was estimated to be responsible for around 24 million metric tons of greenhouse gases and 950 million pounds of hazardous materials, making it the country's number one producer of toxic waste.[32]

In 2011 Koch Industries spent around $8 million on lobbying Congress with a specific focus on environmental issues. When the 112th Congress opened, the House Energy and Commerce Committee was made up of members who have advocated in favour of the oil industry. Twenty-two out of 31 Republicans and five Democratic members had received funding from Koch Industries which had outspent even ExxonMobil. The results favoured the Kochs' agenda. Most of the members of the Committee signed up to the "No Climate Tax" pledge (around 200 elected officials at all levels had signed up) promoted by Americans for Prosperity (also funded by the Kochs). Some members such as Morgan Griffith openly criticised climate science. Griffith then went on to lead a "war on the EPA" within the House which led to a 16% reduction in the agency's budget. Mayer argues this reduced the ability of the EPA to enforce environmental regulations on areas such as toxic pollution which materially benefited Koch Industries.

In the UK the political influence of the *polluter elite* is not as clear-cut compared to the influence they have on the current Trump administration. However, there is some evidence of the political influence of the *polluter elite* such as continued budget support for the oil and gas sector. In recent years the government has regularly put in place tax breaks and other measures such as seismic surveys for drilling for the sector. In 2015 and 2016 this fiscal support was estimated to be worth £10.16 billion.[33] In 2017 the government announced tax relief worth £24 billion for companies to decommission oil and gas platforms in the North Sea.[34] A European Commission study published earlier this year found that the UK had the largest subsidies in Europe, worth around €12 billion for fossil fuels compared to €8.3 billion for renewable energy.[35] These tax breaks are a result of effective lobbying of the government through meetings, in particular with the Treasury, and successfully shaping the narrative around the North Sea oil and gas as an issue of protecting hundreds of thousands of jobs and saving companies that are facing collapse in a difficult economic environment.[36] The UK government also uses billions of pounds of public funding to support oil and gas projects overseas.[37]

The case of billionaire Sir Ian Wood is an emblematic example of the access of the *polluter elite* and their ability to directly influence legislation that favours fossil fuels. He built his fortune at the Wood Group which expanded from providing services to oil and gas operations in the North Sea to operating globally (today the Wood Group also works on offshore and onshore wind). In 2006 he donated £22,000 to the Conservative party.[38] In 2013 the Conservative-Liberal Democrat Coalition asked him to "conduct an independently led review of UK continental shelf oil and gas recovery, specifically looking at how economic recovery could be maximised".[39] It is reasonable to expect that he was chosen to lead this work because of his expertise. When Wood retired from the company on 31 October 2012 he held 21,473,637 shares[40] which at the time were worth £182,311,178.[41,42] It is reasonable to assume that when he did the review and in subsequent years he continued to hold some of these shares. The government has followed all of the Wood Review's recommendations including setting up the Oil and Gas Authority to "regulate, influence and promote the UK offshore oil and gas industry in order to achieve the statutory principal objective of maximising the economic recovery of the UK's oil and gas resources". This was enshrined in two key pieces of legislation, the Infrastructure Act approved in 2015 and the Energy Act approved in 2016. Since its creation the Oil and Gas Authority has facilitated new extraction of oil and gas by spending close to £20 million on seismic exploration, and publicly advocated the viability of extraction at the Clair Ridge field and the potential for the Glengorm field.[43]

Another emblematic example of the political power of the *polluter elite* in the UK is the Conservative government's strong backing for the nascent fracking industry despite opposition from local councils and communities. This support has included personal declarations of support from previous Prime Minister David Cameron[44] and Chancellor George Osborne, who also wrote to his fellow MPs to personally ask them to push for fracking.[45] The Conservatives have also smoothed planning rules and provided financial support.[46] Overall, fracking has been identified as a clear example of the *polluter elite* resisting a shift away from fossil fuels.[47] The politics of fracking will be covered in more detail below including how protests have slowed it down.

BOX 4.2 *POLLUTER ELITE* PROFILE: JIM RATCLIFFE

In 2018 Jim Ratcliffe's estimated net worth was £21 billion making him the richest person in Britain. His wealth is largely based on his 61.8% shareholding in INEOS, the petrochemical company he founded, which produces the raw materials to make products such as plastics (e.g. for packaging), medicines and car parts.[48] INEOS has not disclosed its greenhouse gas emissions. Campaigners argue INEOS has a long track record of pollution, including chemical fires and air pollution. The Grangemouth refinery is reportedly one of the largest emitters of carbon dioxide in Scotland.[49] In 2016 the company completed a £200 million shale gas tank at Grangemouth to store imported fracked gas from the US.

The company wants to see a huge expansion of fracking in the UK as it has built up over a million acres in exploration rights. However, to date this has not been possible. On the day the Scottish government announced a moratorium on granting planning permission for fracking pending consultation and impact assessment, Ratcliffe met with Scotland's First Minister, Nicola Sturgeon.[50] The outcome of the meeting is unknown. INEOS also went through the courts to attempt to overturn the "effective ban" on fracking in Scotland but lost in the summer of 2018.

In February 2019 Ratcliffe criticised the limit on seismic activity saying:

> We have a non-existent energy strategy and are heading towards an energy crisis that will do long term and irreparable damage to the economy and the government needs to decide whether they are finally going to put the country first and develop a workable UK onshore gas industry.[51]

This was after another fracking company Cuadrilla was forced to stop fracking in Lancashire at the end of 2018 because it breached the existing limit of 0.5 magnitude seismic activity four times with tremors of 1.1 and 0.8 magnitude registered.

INEOS was in discussion with ConocoPhillips about buying around £2 billion worth of the company's assets in the North Sea; by February 2019 they had not reached a deal. If the deal were to go through this would be an example of Ratcliffe potentially profiting further from opportunities in the fossil fuel sector.

At the international level several fossil fuel companies have been successfully influencing the UN talks since the first meeting in 1994. At that meeting in Berlin the International Climate Change Partnership which brought together companies and trade associations provided a draft text for the US delegation to use which ended up in the final decision document.[52] This stated that policy should continue to be made at the national level which favoured national level lobbying. Crucially, these industry groups lobbied prior to the actual conference before national delegations arrived at the conferences. The latest example of this lobbying is that at the last two climate change conferences at the end of 2017 and 2018 the US delegation has held public events advocating the use of fossil fuels.

One reason for slow progress at the negotiations is the successful effort by polluting companies to lobby for delay or weak market mechanisms that do not threaten their overall business model and indeed are part of the goal of increasing economic growth.[53] Several of the companies listed in the *polluter elite* database (including BP, BHP Billiton, Chevron, Rio Tinto and Shell) are members of the International Emissions Trading Association which has a long track record of lobbying at the UN climate change talks to get their priorities into the negotiation text.[54]

BOX 4.3 THE GLOBAL *POLLUTER ELITE*

This book has focused on privately owned fossil fuel companies. But state companies are responsible for a huge share of historical greenhouse gas emissions, and possess the majority of fossil fuel reserves.[55] Looking at the people who run both types of companies a future research agenda could build up an understanding of this global *polluter elite*. It would be crucial to focus in on the decision makers who run these firms because in each country they have political influence over the government in deciding the level of production and consumption from fossil fuels *or* clean alternatives (both state[56] and privately owned companies pressure the government to secure fossil fuel subsidies). These decisions in turn heavily determine the consumption options available to the wider population. This research agenda could focus on the individuals running the major fossil fuel producers who comprise the Organization of the Petroleum Exporting Countries (OPEC), such as Saudi Arabia and Venezuela, and major consumers such as China and India.

Research on the global *polluter elite* and their common interests (as Chapter 1 highlighted the wealthiest people are hypermobile leading some to see them as stateless) could be crucial to further understand the continuing inadequate progress at the UN climate change negotiations. Current voluntary pledges under the Paris Agreement of 2015 are not ambitious enough to keep temperature rise below 1.5 degrees Celsius, let alone 2 degrees,[57] and there has been very limited progress since 2015.

Studying the global *polluter elite* would enable further analysis to look beyond surface differences (e.g. by nationality and negotiating bloc) to see how the global *polluter elite* use their political influence to reinforce the fossil fuel status quo in each country, which then influences that country's negotiating position at the UNFCCC (e.g. Saudi Arabia, Kuwait, Russia and the US weakened acceptance of the 2018 IPCC report at the UN climate change conference in Poland in December 2018).[58] This could help explain why progress since 1994 has been glacially slow despite climate science becoming much clearer through successive IPCC reports.

In the past UN climate change talks the negotiators for countries such as China and India have been accused of hiding behind their huge populations of poorer citizens to justify their argument that countries in the global south need "space to grow" and therefore emit. A study on the global *polluter elite* would be in this spirit but would apply across all countries whether in the global north or global south. This could look at how the richest people in each country, and globally (as they invest in polluting operations globally), hide behind the general population. This is by using arguments that the general population is addicted to fossil fuels to justify the continued use of fossil fuels as there are no "viable alternatives" to "keep the lights on" and provide the energy needed for conventional economic growth. In effect the *polluter elite* in different countries agree on the continued use of fossil fuels and the deadlock at the UNFCCC suits them. While the global *polluter elite* have a common interest in undermining global and national efforts to reduce emissions it is unlikely there is a completely coordinated shared agenda as they also compete with each other. This research agenda could examine how the global *polluter elite* interact, for example, if there is a hierarchy of influence and unwritten rules about which members of this group influence others.

4.4 The political influence of the *polluter elite* has been decisive because they exert structural power over the state

The *polluter elite* do not exercise their political influence in a vacuum. To deepen understanding of why the political influence of the *polluter elite* has been so effective to date requires placing their role in context. Given the deep links between incumbent power and the state this must include a nuanced understanding of the state (see Box 4.4). Despite the state being referred to frequently in the literature on sustainability transitions it is rarely examined in detail. This means moving beyond seeing the state as a neutral manager. The reality is that the policy-makers running the state interact with and, as will be discussed below, to a certain extent rely on the *polluter elite*. There can be no assumption that policymakers are neutral and inherently want to accelerate the low-carbon transition.

In looking at the role of the *polluter elite* this requires understanding of how they interact with the state because as Mills put it in reference to his concept of the power elite,

> if our interest in the very rich goes beyond their lavish or their miserly consumption, we must examine their relations to modern forms of corporate property as well as to the state: for such relations now determine the chance of men to secure big property and to receive high income.[59]

BOX 4.4 THE ROLE OF THE STATE IN SUSTAINABILITY TRANSITIONS (BASED ON JOHNSTONE AND NEWELL, 2018).[60]

The state is not neutral

The state should not be treated as an independent and atomised actor. State policy leads to winners and losers. This is highlighted by the structural power of the *polluter elite* who have formed alliances with the state, in particular when the Republicans and Conservatives are in government.

There are multiple dimensions of state power

The state is not a monolithic entity. This is illustrated by contradictions between different parts of the state apparatus. There are ministries supporting the expansion of fossil fuel production (subsidies via the finance ministry) while in theory environment ministries seek to protect the natural world. This is often why the *polluter elite* seek to influence the more powerful ministries of finance, trade and industry. Another example of these contradictions is that both the US and the UK have institutions working on climate science (e.g. NASA and the Climate Change Committee) whilst the military is one of the largest users of oil.

A further contradiction can be between national policy and that of the local government; whilst the focus of this book is on the national government, in both countries the sub-national levels (e.g. Federal states, local councils, etc) have taken different positions on climate change. For example, the state of California has committed to 100% renewable energy by 2045.[61] In the UK several local councils have denied planning permission for fracking. The Scottish government has an effective moratorium on fracking, whilst the UK government promotes fracking in England.

The state is influenced by the dominant ideology of the day

Since the 1980s neo-liberalism has entrenched itself. This ideology is based on the viewpoint that the state should be small and should facilitate free markets, rather than taking an active role in the economy. These beliefs have underpinned state policy in the US and the UK over the last 40 years which has privatised state companies (energy, water, transport, etc), deregulated the financial sector and reduced taxes on corporations and the wealthiest.

The *polluter elite* have been direct beneficiaries of these policies, which have contributed to increasing their wealth (the elite have paid less tax and are able to make money from speculative investments in the financial sector) whilst simultaneously weakening the ability of the state to proactively lead the low-carbon transition.

The central goal of the state in the US and the UK is economic growth because this is the basis of its legitimacy. Governments have been obsessed with economic growth since Simon Kuznets developed a way to measure the Gross Domestic Product (GDP). He did this to measure annual economic activity for the US national accounts in the 1930s. Even though GDP has been heavily questioned as a measure of progress,[62] it is still what the largest political parties in the US (Republican and Democrat) and the UK (Conservative, Labour, Liberal Democrats) advocate. Since the Second World War the "economy" has often been a central issue in US and UK elections and comes high on survey respondents' list of priorities. This means debates end up being focused on the record of GDP growth or decline, with policymakers seeing a strong link to the availability of energy.

The goal of increasing economic growth creates a type of dependency of the state on those existing industries that already provide jobs and revenue for the government, and in particular those companies that generate energy to power economic activity. As a result of this dependency the fossil fuel sector has ended up having more "structural power" over state decision-making (Newell and Paterson, 1998)[63] because this sector has had higher rates of profitability and reinvestment, as well as innovations in production. This has been the case for some time and perhaps helps to explain the consistent state support for fossil fuels over such a long period. This structural power is so effective because successive governments have inherited a deep dependency on continued growth (capital accumulation) in certain economic sectors to secure their legitimacy.[64] The other side of this scenario is that broadly speaking the state has not trusted renewable energy to be a new site of capital accumulation.[65]

The automobile industry, which has strong links to fossil fuels and is a vital manufacturing sector, has used its crucial position in the economy to influence government decisions on regulations and direct support at times of crisis.[66] The fear of national and local government of the economic consequences of a car factory closing can add to the pressure to ensure the industry continues. This vulnerability has been particularly exposed during the prolonged Brexit negotiations where a number of vehicle manufacturers have threatened to, or actually relocated from the UK.

Such companies exert structural power over the state because they are able to threaten reducing investment (known as the "investment strike")[67] or relocating operations, both of which could undermine the state's legitimacy which rests on rising economic growth and jobs (Figure 4.3). The *polluter elite* who run these companies have made it clear to policymakers that they will not tolerate restrictions on emissions which would mean drastic reductions in production and consumption. For example, George W. Bush justified withdrawal from the Kyoto Protocol in 2001 because of the perceived danger of companies outsourcing their operations from the US due to limits on emissions. In a globalised economy where capital can flow freely this is a particularly effective tactic that companies can use. This can lead to rising emissions as carbon-intensive operations constantly relocate to where production costs such as labour and electricity are lowest facilitated by governments that are willing to build fossil fuel energy infrastructure.[68]

The dependency of the state on fossil fuels is a key factor in their decision to continue to support this sector (Figure 4.3). The incumbent

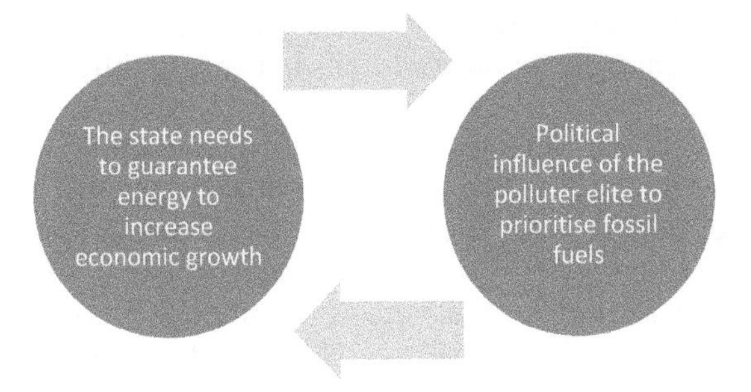

Figure 4.3 The basis of the alliance between the polluter elite and policymakers.
Source: Author.

regime is generally seen to be made up of polluting companies and policymakers. It is this "core regime level alliance" that works to defend the status quo (Geels, 2014).[69] This alliance explains why governments in the US and the UK, as well as elsewhere, have failed to proactively accelerate low-carbon transitions and instead continue to back fossil fuels.[70] It becomes clear why the policymakers have prioritised the interests of the *polluter elite* ahead of coalitions who want action on climate change. The dependency of the state leads to close contact between the *polluter elite* and policymakers (what Geels refers to as relational networks).[71]

In this context of alliances being formed it is key to understand that the lobbying by the fossil fuel industry is not a one-way process. One indicator of this is that in the US in recent years around 50% of retiring senators and an estimated 42% of retiring representatives have become lobbyists regardless of party affiliation.[72] The revolving door between companies and public office blurs the divide between the interests of a private company and state policy. In effect the state is not operating in a separate space from the fossil fuel industry. Different groups revolve around and within the state through a range of networks.[73] An emblematic case of this is when Rex Tillerson moved from being CEO of ExxonMobil to Secretary of State under the Trump administration.

In the UK the goal has been to capture regulatory institutions to ensure individuals friendly to the fossil fuel sector are in these positions.[74] Examples include a member of the executive team of a fracking company working in the Cabinet Office, and investors and lobbyists in fracking companies filling the role of energy advisor and chief scientific advisor to the government. Instead of being neutral these policymakers are part of "deep incumbency". These examples of the revolving door, that have even been discussed in Parliament, demonstrate blurred lines between public interest and private companies.

Such overlaps and alliances (whether formal or informal) help explain why the US and UK state have continued to subsidise the fossil fuel sector despite the growing scientific evidence of global warming. The structural power of the *polluter elite* also explains contradictions between different parts of the state (as highlighted above it is not a monolith), with scientific bodies (e.g. NASA and the Climate Change Committee) warning of the dangers of climate change while the ministries overseeing energy are promoting increased extraction of fossil fuels.

4.4.1 The structural power of the **polluter elite** *and the energy trilemma*

An example of this alliance in action is when governments use justifications for continued use of fossil fuels which repeat key messaging from the fossil fuel industry. This section will show the *polluter elite* have effectively pressured governments to slow down the transition by highlighting the perceived costs of climate change policies. This strategy has used the logic of the energy trilemma concept which sees the three dimensions of energy security, reducing emissions and avoiding negative social impact as conflicting goals (Figure 4.4).[75]

Despite the breadth of their resources, the *polluter elite* have not always been able to influence government policy to deepen the fossil fuel economy. There are moments since 1990 when the US and the UK government have introduced measures that had the potential to reduce greenhouse gas emissions and directly threaten the interests of the *polluter elite* (see Appendix 2). For example, the Obama administration introduced the Clean Power Plan via the EPA in August 2015. This was a standard that aimed to reduce emissions from electricity generation by around 32% by 2030 (based on 2005 levels). President Obama also played a leading role in ensuring the Paris Agreement was reached in December 2015. In 2016 he ratified the agreement. Meanwhile, in the UK the Labour government pushed forward the approval of the Climate Change Act in 2008 (legally binding carbon budgets to reduce emissions by 80% by 2050) and introduced a Feed-in Tariff in 2010 so that any producers of solar power (including households) could sell power back to the grid for a guaranteed fee.

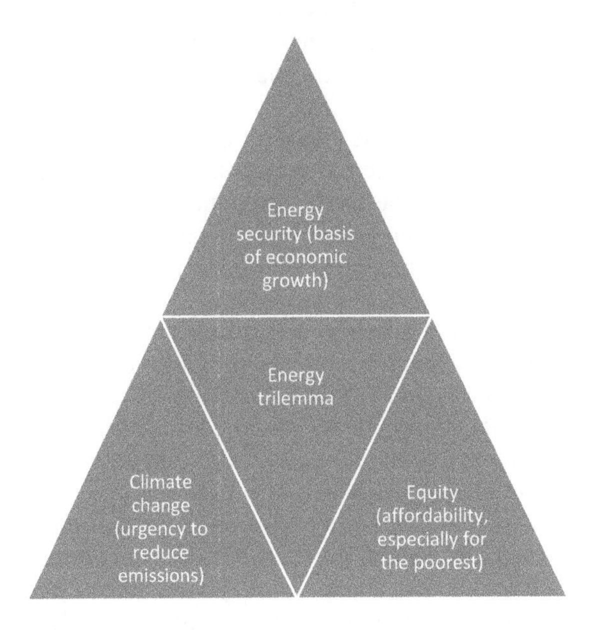

Figure 4.4 The conflict between the different dimensions of the energy trilemma.
Source: Author.

Two key factors help to explain why the Obama and Labour governments went against the structural power of the *polluter elite*:

- Substantial pressure from civil society

  In September 2014 around 300,000 people marched through New York in the largest ever march on climate change. Meanwhile, the Clean Power Plan received a record number of supportive comments from a wide range of US citizens from sectors such as health, trade unions, religious groups, youth groups and business.[76]

  In the UK civil society organised from 2005 to 2008 to pressure for the introduction and subsequent approval of the Climate Change Act. This included 200,000 people contacting their MPs to ask them to support the legislation. This pressure was also fundamental in increasing the ambition of the Act which just two weeks before it was voted on increased the emissions reduction target from 60% to 80%.[77] A coalition of civil society groups (including the Renewable Energy Association, Friends of the Earth, Coop Bank, National Farmers Union) pressured the Labour government to approve the Energy Act in 2010 which included the Feed-in Tariff (a subsidy to pay solar installations for selling excess electricity back to the grid).

- Politicians were able to justify these policies as greener ways to increase economic growth.

When they passed these measures both governments promised they would lead to economic growth and reduce greenhouse gas emissions. In effect the Democrats and New Labour accepted the climate science that global warming had to be addressed and looked to reconcile this goal with continued economic growth, often based on a vision of free markets.[78] Their actions were coherent with the state's need to ensure economic growth for its legitimacy. When Obama announced the Clean Power Plan, he said a transition to clean energy would create jobs and cited statistics showing that the solar industry was already creating jobs ten times faster than the rest of the economy.[79] Ed Miliband was the Energy and Climate Change Minister who played a crucial role in the passage of the Climate Change Act. When it was announced, he said it was part of a set of policies that were fair, sustainable and "which meets our obligations to today's and future generations". He argued the Stern Report demonstrated it would be costlier to do nothing. In his speech to the Trades Unions Congress in 2009 the Miliband said: "We also need to learn the lesson that climate change is no longer just about the environment. It has got to be about jobs, energy security and fairness as well".[80]

However, the *polluter elite* in alliance with Republican and Conservative policymakers have invoked the energy trilemma to justify undermining climate change measures introduced by the Democrats and the Labour party. In the US President Trump, himself a member of the *polluter elite*, invoked energy security for economic growth as the justification to withdraw from the Paris Agreement in 2017. He said:

> As President, I can put no other consideration before the well-being of American citizens. The Paris Climate Accord is simply the latest example of Washington entering into an agreement that disadvantages the United States to the exclusive benefit of other countries, leaving American workers — who I love — and taxpayers to absorb the cost in terms of lost jobs, lower wages, shuttered factories, and vastly diminished economic production.[81]

In an example of history repeating itself very similar arguments were used by President George. W. Bush to withdraw from the Kyoto Protocol in 2001.

Since President Trump has been in office he has also instructed the EPA to roll back the Clean Power Plan. In early 2019 the EPA announced it was proposing the Affordable Clean Energy rule which relies on efficiency improvements at power plants with no specific emissions reduction targets and could see emissions *increase* at 28% of power plants. The justification for this was to ensure the US economy remains competitive (energy security dimension) and reduce electricity costs for poor Americans (equity dimension).

In the US the *polluter elite* have also used their superior resources to achieve substantial success in questioning the urgency to reduce emissions (one dimension of the energy trilemma) by funding and advocating climate change denial. A dossier compiled by the Union of Concerned Scientists claims that BP, Chevron, ConocoPhillips, ExxonMobil, Peabody Energy, and Shell knew about climate change decades ago and approved funding for climate change denial to confuse the science.[82] The Koch brothers have also catalysed large individual donations. This has had a huge impact on the Republican party, leading Mayer to argue it has mostly adopted the Koch brothers' personal positions on climate change and tax policy.[83] Democrat Senator Sheldon Whitehouse has stated that "talking to Republicans about climate change is like talking to prisoners about escape. They may want out, but they can't have their fossil-fuel wardens find out".[84] According to Brulle a range of research shows that one major factor driving this misunderstanding is a "climate change counter-movement". This is a deliberate and organized effort to misdirect the public discussion and distort the public's understanding of climate change. Conservative think tanks, trade associations and advocacy groups have played a major role in successfully delaying meaningful government policy actions to address the issue, such as when Republican politicians blocked cap-and-trade legislation in 2010 that would have restricted emissions.[85]

In the UK the Conservative government has used the energy security and equity dimensions of the energy trilemma since 2010 to promote oil and gas whilst simultaneously undermining the low-carbon economy. Echoing statements by companies such as Cuadrilla, Chancellor George Osborne claimed in 2013 that fracking had the potential to create jobs and bring down energy bills.[86] This was one of the justifications for announcing a package of measures to support fracking in 2017. Minister of State for Energy and Clean Growth Claire Perry said: "British shale gas has the potential to help lower bills and increase the security of the UK's energy supply while creating high quality jobs in a cutting-edge sector".[87]

Despite almost all Conservative MPs voting in favour of the Climate Change Act in 2008, once the Conservative party was elected in 2010, Chancellor George Osborne argued the UK should not reduce its competitiveness by pursuing green policies that would "pile costs on the energy bills of households and companies". In November 2013 Prime Minister David Cameron reportedly said he wanted to "get rid of all the 'green crap'". The pinnacle of this rhetoric came in 2015 when the Conservative government cut a wide range of support because it said it had raised consumers' energy bills.[88] This included cuts in subsidies for solar, biomass, and biogas, as well as scrapping support for energy efficiency in households and offices. In direct contradiction to the Climate Change Act all of these measures were estimated to increase the UK's emissions.[89] They also had a devastating impact on the renewable energy sector because the removal of the subsidies undermined the commercial viability of solar companies. An estimated 12,000 jobs were lost in the year following the cuts, and a 95% fall in investments was forecast.[90] One of the most symbolic measures was to cut the Feed-In-Tariff (FiT) which had been successful in increasing renewable energy supply since it was introduced in 2010. Even though pressure from civil society and business was crucial to the FiT being introduced in the first place and continuing despite government attempts to undermine it, such pressure was ultimately unsuccessful. In December 2018 the government announced it was ending the FiT for all new installations. Thus, any installation set up after 31 March 2019 will transfer unused solar power back to the national grid for free.[91]

These decisions by the Conservatives have been especially damaging to the low-carbon transition because they simultaneously undermine emerging green innovations and reinforce the role of oil and gas in the economy. In effect what happened is that the potential of initiatives to disrupt carbon lock-in and catalyse decarbonisation pathways was intentionally reduced to ensure the supremacy of fossil fuels. This overall process of undermining renewable energy whilst supporting fracking (and nuclear) has been described as "destructive recreation" because instead of green niches emerging incumbents were reinforced, meaning they avoided a process of "creative destruction" (Johnstone et al, 2017).[92] The Conservative government did launch a Clean Growth Strategy in 2017 to encourage emissions reductions across transport, electricity generation and buildings (including committing £2.5 billion for low-carbon innovation).[93] But overall, the way in which this same government has undermined the Climate Change Act by supporting increase of fossil fuels is a reminder that the approval of such ambitious legislation is the beginning of a political struggle that can last for several decades. The Act has not prevented the Conservative government's efforts to expand fossil fuel extraction in the North Sea via the Oil and Gas Authority and via fracking.

In summary, it is the structural power of the *polluter elite* and the significant resources that they invest in lobbying that are slowing down the transition away from fossil fuels in the US and the UK.

BOX 4.5 THE HISTORIC *POLLUTER ELITE*

The current political activity of the *polluter elite* is not a new trend. The owners of fossil fuel assets have long sought to capture the state to defend their investments and to open up new opportunities. In the US in 1913 the oil industry successfully pressured for a special tax loophole, justified on the grounds that oil exploration was costly, which in practice allowed it to evade tax. The loophole was enlarged in 1926 and it took another 50 years before it was closed.[94] During this time, the oil industry received a range of benefits including tax breaks, large government contracts and support with pipelines. Domhoff claims a coalition of private oil companies lobbied Congress to stop the creation of a state oil company in the 1940s that would have bought oil directly from Saudi Arabia.[95] A startling example of the historical responsibility that the US *polluter elite* hold for climate change is how the Rockefeller family lobbied Congress to approve the Marshall Plan from which they profited hugely.[96] Angus argues that the goal of the Plan was to strengthen US companies, in particular oil companies, and this is why most of the funding (around $130 billion in today's money) to Europe had to be used to purchase goods and services from US companies.[97] Between 1948 and 1951 around 10% of Marshall Plan funds were spent on oil which was hugely profitable for oil companies and contributed to the oil addiction in European economies that continues until this day.

Mills argues that without the support of the government, rich individuals and their families in the US would not have become as rich as they are. He argues that much more important than the personal characteristics of men such as Andrew Carnegie (steel) and John D. Rockefeller (oil) are the factors such as the "shape of the economy", the location of resources such as oil, the huge levels of corruption of the political system, legal frameworks, the structure of taxation and tax loopholes for oil.[98] He gives the example of the government paying for the paved road system without which Henry Ford would not have become a billionaire from manufacturing vehicles. Phillips found an ongoing correlation between support from government and the concentration of wealth in certain families, sectors and regions for more than two centuries. He argues that from the beginning of the US republic the government played a "critical, sometimes brazen"

(Continued)

role in creating economic elites through political favours such as grants, tariff protection, tax advantages and monopolies.[99]

In the UK coal owners coordinated to ensure a favourable business environment. Jaffe argues the coal owners joined forces between 1800 and 1840 to form a cartel to ensure favourable prices, even if there was not always unity within this group.[100] A Parliamentary select committee was appointed in 1836 to investigate the claim of London residents that a coal cartel was preventing competition and keeping prices high. Several coal owners including Hedworth Lambton, Joseph Pease and William Bell were able to defend the coal cartel. Based on his collieries around Bishop Auckland, Pease was one of the biggest coal and coke producers in the North-east throughout the mid-1800s.[101] James Losh, a coal owner in northern England, describes in his 1824–1833 diaries how he would meet with other colliery (coal pit) owners to discuss how to resist government measures which would have reduced coal prices. These other coal owners included the Marquis of Londonderry, Lord Wharncliffe, Lord Ravensworth, Mr. Russell, Captain Cochrane, Mr. Bell and Mr. Liddle.[102] In 1854 the coal owners formed the Mining Association of Great Britain (MAGB.) In a context of growing demands for nationalisation the MAGB intensively and successfully lobbied Tory MPs to halt nationalisation immediately following the First World War. Through its privileged access to Ministers the MAGB was able to successfully water down legislation in 1920, 1929–30, 1936–8 and also managed to obtain subsidies in 1921 and 1925–1926. However, they were unable to prevent the coal industry from being nationalised in 1947 after decades of tensions between owners and miners.

4.5 Increasing economic inequality is reinforcing the political power and resources of the *polluter elite*

The *polluter elite* will never stop seeking political influence. As the fossil fuel incumbents, they have too much to lose by allowing the acceleration of the low-carbon transition to happen. The issue is that at a time of rising inequality the *polluter elite* are amassing increasing resources for lobbying. This also strengthens their ability to exert structural power over the state. This highlights that the status of incumbents is dynamic. They can become more powerful regardless of whether niches such as renewable energy are also growing in size. It is important to factor in inequality when analysing the contested transition in the US and the UK because if the richest people are able to capture more resources this tends to lead to more influence over government policy.[103] Lower taxes and access to tax havens mean the people who are part of the incumbent regime are increasing their resources at precisely the time when the *polluter elite* needs to be destabilised to accelerate the transition away from fossil fuels.

The *polluter elite* and the companies they run are benefiting from lower taxes. This trend began under the neo-liberal policies of President Ronald Reagan and Prime Minister Margaret Thatcher in the 1980s. In the US the top income tax rate dropped from 70% in the early 1980s to 40% in 2013. On average the top tax rate was 39% between 1980 and 2009 compared to an average of 75% between 1950 and 1979.[104] This led Piketty to ask "has the US political process been captured by the 1 percent?".[105] President Trump approved a further substantial tax cut in early 2018 that also saw corporation tax fall from 35% to 21% which further increases the resources available to the oil and gas companies run by the *polluter elite.*

Likewise, inequality has grown in the UK since the 1980s as changes were made to reduce taxes and transfers. Inheritance tax fell from 75% in 1970 to 40% in 2013. Top rate income taxes fell from 98% in 1970 to 45% in 2013. The Conservative government that has been in power since 2010 has consistently cut taxes for the wealthy. In April 2017 the Resolution Foundation estimated that £2 billion worth of income tax cuts, alongside austerity policies to cut welfare totalling over £1 billion, were in effect a large transfer of wealth from the middle and working class to the rich.[106] The Conservatives have also reduced corporation tax from 28% in 2010 to 19% in 2018[107] which further increases the resources available to the oil and gas companies run by the *polluter elite.*

Another way in which the *polluter elite* and the companies they run have increased resources at their disposal is by using tax havens. According to data brought together by the International Consortium of Investigative Journalists several of the companies and individual members of the *polluter elite* have used tax havens. Their database combines the Panama Papers, Paradise Papers, Offshore Leaks and Bahamas Leaks which can be used to search for the names of companies and individuals. According to data from the London Stock Exchange in 2015 several of the listed oil and gas companies are incorporated in tax havens.[108]

In the US and the UK there are tax rules that facilitate a lower payment of tax through legal offshore tax havens (tax avoidance) and a lack of enforcement for taxes that have not been paid (tax evasion). Harrington argues these structures are created by governments in order for the richest people to become richer as they are able to bypass legal constraints on tax, currency controls and other regulations.[109] Tax havens are estimated to have a third of the world's trade, around US$21–32 trillion, pass through them. Zucman calculated that around $7.6 trillion of household financial wealth (bank deposits and portfolios net of any debt) are held in offshore tax havens.[110] In the US he calculated that around $1.2 trillion was held offshore in 2014 resulting in the US government losing out on around $35 billion in tax revenue. The leaked Panama and Paradise papers have sparked moral outrage with Harrington observing that these leaks have brought us to "stage 1 of the informal social punishment process: naming and shaming".

By paying less tax and using tax havens the *polluter elite*, as part of the broader group of the richest people in the US and UK, are capturing more and more income and wealth. As previously mentioned, this has led some observers to argue there is a trend of plutonomy in countries such as the US and UK.[111] This is where the richest are the group that holds the majority of wealth, receives the highest income and consumes more than other groups (Figures 4.5 and 4.6).

The consequences of extreme economic inequality further concentrating wealth in the hands of the *polluter elite* are particularly striking in the US. The *polluter elite* are using their superior resources to influence policy to favour fossil fuels. Billionaires and millionaires with investments in fossil fuels are donating huge sums to political parties. The following selected quotes from a speech by Senator Sheldon Whitehouse in May 2018 provide a striking example of an elected politician revealing the negative impact of this cycle of economic and political inequality:[112]

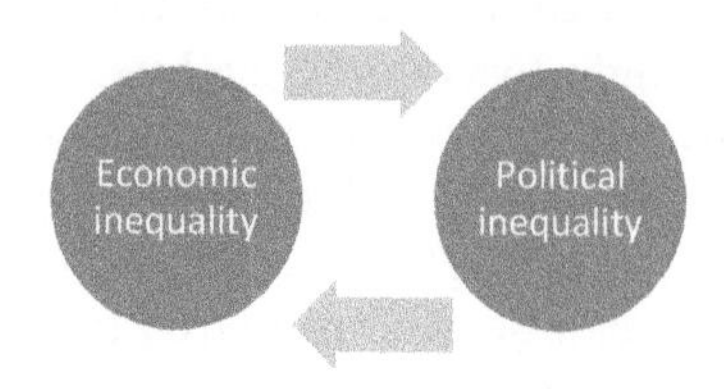

Figure 4.5 Economic inequality and political inequality reinforce each other.
Source: Author.

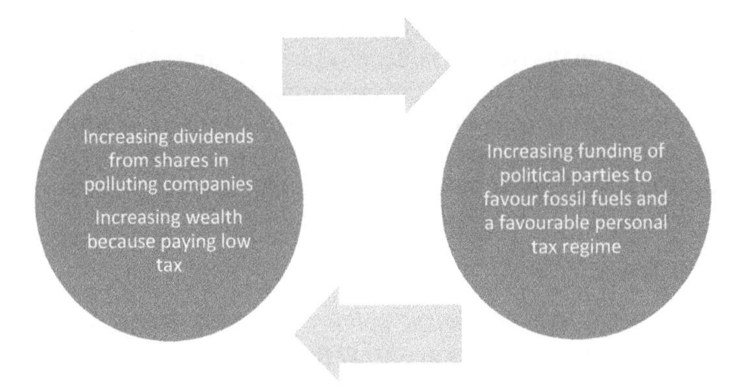

Figure 4.6 The reinforcing cycle of polluter elite profit and political influence.
Source: Author.

> "*What stymies Congress from legislating, or even having hearings, about climate change? The fossil fuel industry determined to control one party on this question; they determined to silence or punish or remove any dissent in one political party.*
>
> *Underneath the illusory democratic surface run subterranean rivers of dark money. Massive infrastructures have been erected to hide that dark-money flow from the sunlight of public scrutiny, to carve out subterranean caverns through which the dark money flows.*
>
> *If you want to understand why we do nothing on climate, you have to look down into those subterranean chambers, understand the dark money, and not be fooled by the surface spectacle.*
>
> *The Senate was legendarily corrupt in the Gilded Age. We cannot pretend it is impossible for the United States Senate to be corrupted; our history refutes that happy thought.*
>
> *Congress remains a democratic body on the surface, with all the procedural veneer and trappings of democracy: we hold votes, and there are caucuses and hearings and voting. But on the issues that most concern the biggest special interests, Congress no longer provides America a functioning democracy.*
>
> *And perhaps the worst of all is that the world is watching. It is watching us as the fossil fuel industry, its creepy billionaires, its front groups, its bogus think tanks, all gang up and debauch our democracy.*"

In relation to environmental policy the level of inequality and asymmetric power have an impact on which policies are put in place, how they affect business and who benefits or loses.[113] Boyce has identified the dynamic whereby those with more money have more political influence. This is when the rich are politically powerful and able to influence the state to prioritise profit-making over public health regardless of how many ordinary citizens would prefer to have a healthy environment. The rich and politically powerful benefit from economic activity that destroys the environment while the poor suffer the consequences.[114]

4.6 Conclusion

This chapter has shown how the *polluter elite* have access to policymakers and have had success in setting the direction of state policy by slowing the low-carbon transition. Such is their political influence that instead of phasing out fossil fuels in response to the increasing scientific evidence of global warming the US and the UK have subsidised extraction of fossil fuels.

The *polluter elite* have used their structural power over the state to ensure that policymakers have justified continued support for fossil fuels based on arguments that it will guarantee energy security (crucial for economic growth) and reduced energy costs for the poor. That said, their political power has not been absolute. Policymakers have implemented legislation to reduce emissions based on a vision of green growth and following pressure by civil society. However, in an example of the *polluter elite*'s continued ability to slow down the low-carbon transition they have strategically overcome opposition by successfully pressuring for the weakening or reversal of these measures. At a time of rising inequality their political power is growing.

Notes

1 Köhler, J. et al., 2019. An Agenda for Sustainability Transitions Research: State of the Art and Future Directions. *Environmental Innovations and Societal Transitions*, [online]. Retrieved from: doi:10.1016/j.eist.2019.01.004 [Accessed 17 February 2019]; Fischer, L. and Newig, J., 2016. Importance of Actors and Agency in Sustainability Transitions: A Systematic Exploration of the Literature. *Sustainability*, [online], 8 (5). Retrieved from: www.mdpi.com/2071-1050/8/5/476 [Accessed 17 February 2019].
2 Geels, F. W. and Schot, J. W., 2007. Typology of Sociotechnical Transition Pathways. *Research Policy*, [online], 36 (3). Retrieved from: doi:10.1016/j.respol.2007.01.003 [Accessed 17 February 2019]; Avelino, F., 2017. Power in Sustainability Transitions: Analysing Power and (dis)empowerment in Transformative Change Towards Sustainability. *Environmental Policy and Governance*, [online], 27 (6). Retrieved from: doi:10.1002/eet.1777 [Accessed 17 December 2018]; Realising Transition Pathways. EPSRC 'Realising Transition Pathways' Consortium:

Key Findings. Retrieved from: www.realisingtransitionpathways.org.uk/realisingtransitionpathways/publications/RTP_Achievements_One_Pager_Final.pdf [Accessed 20 August 2018].

3 Köhler, J. et al., 2019. An Agenda for Sustainability Transitions Research: State of the Art and Future Directions. *Environmental Innovations and Societal Transitions*, [online]. Retrieved from: doi:10.1016/j.eist.2019.01.004 [Accessed 17 February 2019].

4 Sovacool, B. K., 2016. How Long Will It Take? Conceptualizing the Temporal Dynamics of Energy Transitions. *Energy Research and Social Science*, [online], 13 (202–215). Retrieved from: doi:10.1016/j.erss.2015.12.020 [Accessed 17 December 2018].

5 Motkya, M., 2018. 2019 Renewable Energy Industry Outlook. *Deloitte*. Retrieved from: www2.deloitte.com/content/dam/Deloitte/us/Documents/energy-resources/us-renewable-energy-outlook-2019.pdf [Accessed 17 February 2019].

6 Department for Business, Energy and Industrial Strategy, 2018. Digest of UK Energy Statistics. Retrieved from: https://assets.publishing.service.gov.uk/government/uploads/system/uploads/attachment_data/file/736152/Ch5.pdf [Accessed 20 December 2018]; Department for Business, Energy and Industrial Strategy, 2018. Digest of UK Energy Statistics. Retrieved from: https://assets.publishing.service.gov.uk/government/uploads/system/uploads/attachment_data/file/736153/Ch6.pdf [Accessed 20 December 2018].

7 Fouquet, R. and Pearson, P. eds., 2012. Special Section: Past and Prospective Energy Transitions – Insights from History. *Energy Policy*, [online], 50 (1–848). Retrieved from: www.sciencedirect.com/journal/energy-policy/vol/50/suppl/C [Accessed 17 December 2018]; Roberts, C. et al., 2018. The Politics of Accelerating Low-carbon Transitions: Towards a New Research Agenda. *Energy Research and Social Science*, [online], 44 (304–311). Retrieved from: doi: 10.1016/j.erss.2018.06.001 [Accessed 17 December 2018].

8 Wells, P. and Nieuwenhuis, P., 2012. Transition Failure: Understanding Continuity in the Automotive Industry. *Technological Forecasting and Social Change*, [online], 79 (9). Retrieved from: doi:10.1016/j.techfore.2012.06.008 [Accessed 17 February 2019].

9 Hess, D. J., 2014. Sustainability Transitions: A Political Coalition Perspective. *Research Policy*, [online], 43 (2). Retrieved from: doi:10.1016/j.respol.2013.10.008 [Accessed 17 December 2018].

10 Geels, F. W., 2014. Regime Resistance against Low-Carbon Transitions: Introducing Politics and Power into the Multi-Level Perspective. *Theory, Culture and Society*, [online], 31 (5). doi:10.1177%2F0263276414531627 [Accessed 17 December 2018].

11 Köhler, J. et al., 2019. An Agenda for Sustainability Transitions Research: State of the Art and Future Directions. *Environmental Innovations and Societal Transitions*, [online]. Retrieved from: doi:10.1016/j.eist.2019.01.004 [Accessed 17 February 2019].

12 Avelino, F. and Rotmanans, J. W., 2009. Power in Transition: An Interdisciplinary Framework to Study Power in Relation to Structural Change. *European Journal of Social Policy*, [online], 12 (4). Retrieved from: doi:10.1177%2F1368431009349830 [Accessed 17 February 2019].

13 Geels, F. W., 2014. Regime Resistance against Low-Carbon Transitions: Introducing Politics and Power into the Multi-Level Perspective. *Theory,*

Culture and Society, [online], 31 (5). doi:10.1177%2F0263276414531627 [Accessed 17 December 2018].

14 Sovacool, B. K., 2016. How Long Will It Take? Conceptualizing the Temporal Dynamics of Energy Transitions. *Energy Research and Social Science*, [online], 13 (202–215). doi:10.1016/j.erss.2015.12.020 [Accessed 17 December 2018].

15 Short, J. R., 2013. Economic Wealth and Political Power in the Second Gilded Age. *In*: Hay, I., ed. *Geographies of the Super-Rich*. Cheltenham: Elgar, 34.

16 Klein, N., 2014. *This Changes Everything: Capitalism vs the Climate*. New York: Allen Lane.

17 Federal Election Commission. Individual Contributions. Retrieved from: www.fec.gov/data/receipts/individual-contributions/?two_year_transaction_period=2016&contributor_name=Kelcy+Warren&min_date=01%2F01%2F2015&max_date=12%2F31%2F2016 [Accessed 20 December 2018]; Federal Election Commission. 58th Presidential Inaugural Committee. Retrieved from: http://docquery.fec.gov/pdf/286/201704180300150286/201704180300150286.pdf [Accessed 20 December 2018].

18 Department for Business, Energy and Industrial Strategy, 2017. New Measures to Back British Shale Gas Exploration. Retrieved from: www.gov.uk/government/news/new-measures-to-back-british-shale-gas-exploration [Accessed 20 December 2018].

19 Geels, F. W., 2014. Regime Resistance against Low-Carbon Transitions: Introducing Politics and Power into the Multi-Level Perspective. *Theory, Culture and Society*, [online], 31 (5). doi:10.1177%2F0263276414531627 [Accessed 17 December 2018]; Newell, P. and Paterson, M., 1998. A Climate for Business: Global Warming, the State and Capital. *Review of International Political Economy*, [online], 5 (4). Retrieved from: www.jstor.org/stable/4177292?seq=1#page_scan_tab_contents [Accessed 17 February 2018].

20 Pirani, S., 2018. *Burning Up: A History of Fossil Fuel Consumption*. London: Verso Books.

21 Pfund, N. and Healey, B., 2011. What Would Jefferson Do? [online]. *DBL Investors*. Retrieved from: http://i.bnet.com/blogs/dbl_energy_subsidies_paper.pdf [Accessed 20 December 2018].

22 Mayer, J., 2016. *Dark Money: The Hidden History of the Billionaires Behind the Rise of the Radical Right*. New York: Anchor Books.

23 Redman, J., 2017. Dirty Energy Dominance: Dependent on Denial. Oil Change International. Retrieved from: http://priceofoil.org/content/uploads/2017/10/OCI_US-Fossil-Fuel-Subs-2015-16_Final_Oct2017.pdf [Accessed 20 December 2018].

24 Guttsman, W. L., 1968. *The British Political Elite*. London: Macgibbon & Kee.

25 Wilson, R., 2017, 2 May. Oil and Gas Interests Filling Donald Trump's Swamp with Cash. *The Center for Public Integrity*. Retrieved from: https://publicintegrity.org/federal-politics/oil-gas-and-coal-interests-filling-donald-trumps-swamp-with-cash/ [Accessed 21 December 2018].

26 Popovich, N. et al., 2018, 28 December. 78 Environmental Rules on the Way Out Under Trump. *New York Times*. Retrieved from: www.nytimes.

com/interactive/2017/10/05/climate/trump-environment-rules-reversed.html [Accessed 20 November 2017].
27 Hasemyer, D., 2017, 5 June. Choke Hold: The Fossil Fuel Industry's Fight against Climate Policy, Science and Clean Energy. *Inside Climate News*. Retrieved from: https://insideclimatenews.org/content/choke-hold [Accessed 20 November 2017].
28 Holden, E., 2018, 26 November. Trump on Own Administration's Climate Report: 'I don't Believe It'. *The Guardian*. Retrieved from: www.theguardian.com/us-news/2018/nov/26/trump-national-climate-assessment-dont-believe [Accessed 21 December 2018].
29 Singer, M., 2018. *Climate Change and Social Inequality: The Health and Social Costs of Global Warming*. London: Routledge.
30 *The Atlantic*. The Atlantic Creates Searchable, Publicly-available Spreadsheet of FEC PDFs. Retrieved from: www.theatlantic.com/press-releases/archive/2017/01/the-atlantic-creates-searchable-publicly-available-spreadsheet-of-fec-pdfs/513935/ [Accessed 17 February 2019]; Aisch, G., 2016, 18 May. What's In Donald Trump's 104-page Financial Disclosure? *New York Times*. Retrieved from: www.nytimes.com/interactive/2016/05/18/us/politics/trump-financial-disclosure.html [Accessed 20 November 2017].
31 Federal Electoral Commission. Online Database Search. Retrieved from: www.fec.gov/data/browse-data/ [Accessed 17 February 2019]; Center for Responsive Politics. Energy/Natural Resources. Retrieved from: www.opensecrets.org/industries/indus.php?Ind=E [Accessed 17 January 2019].
32 Mayer, J., 2016. *Dark Money: The Hidden History of the Billionaires Behind the Rise of the Radical Right*. New York: Anchor Books.
33 Whitley, S. et al., 2018. G7 Fossil Fuel Subsidy Scorecard. ODI Policy Brief. Retrieved from: G7 fossil fuel subsidy scorecard [Accessed 17 February 2019]; Oil and Fossil Fuel Production (including Gas) Sector Report., 2017. Parliament. Retrieved from: www.parliament.uk/documents/commons-committees/Exiting-the-European-Union/17-19/Sectoral%20Analyses/25-Oil-and-Fossil-Fuel-Production-Report.pdf [Accessed 17 February 2019].
34 Hinson, S., 2018. Future of the UK oil and Gas Industry. *Parliament*. Retrieved from: http://researchbriefings.files.parliament.uk/documents/CDP-2018-0210/CDP-2018-0210.pdf [Accessed 17 February 2019]; HMRC. Statistics of Government revenues from UK Oil and Gas production. Parliament. Retrieved from: https://assets.publishing.service.gov.uk/government/uploads/system/uploads/attachment_data/file/740258/Statistics_of_government_revenues_from_UK_oil_and_gas_production__Sept_2018_.pdf [Accessed 17 February 2019].
35 European Commission. Energy Prices and Costs in Europe. *EC*. Retrieved from: https://eur-lex.europa.eu/legal-content/EN/TXT/PDF/?uri=COM:2019:1:FIN&from=EN [Accessed 17 February 2019].
36 Influence Map. North Sea Oil and Gas Taxation. Retrieved from: https://influencemap.org/site/data/000/257/North_Sea_Oil_and_Gas_Taxation_Lobbying.pdf [Accessed 17 February 2019].
37 Vaughan, A., 2019, 24 February. Ban Ki-moon tells Britain: Stop Investing in Fossil Fuels Overseas. *The Guardian*. Retrieved from: www.theguardian.com/environment/2019/feb/24/ban-ki-moon-britain-stop-invest-fossil-fuels-overseas [Accessed 9 March 2019].

38 The Electoral Commission. Online Database Search. Retrieved from: http://search.electoralcommission.org.uk/ [Accessed 17 February 2019].
39 Wood Review Implementation. Retrieved from: www.gov.uk/government/groups/wood-review-implementation-team [Accessed 17 February 2019].
40 Wood Group. Annual Report and Annual Accounts 2012, 48. Retrieved from: www.woodgroup.com/__data/assets/pdf_file/0015/2427/John-Wood-Group-Plc-Annual-Report-and-Accounts-2012.pdf [Accessed 17 February 2019].
41 Wood plc. Share Performance. Retrieved from: www.woodplc.com/investors/share-performance [Accessed 17 February 2019].
42 Shares multiplied by share price on 31 October 2012 of £8.49; if he still held the same amount on 31 January 2019 these shares would be worth closer to £110 million.
43 Oil & Gas Authority, 2018. Oil & Gas Authority Annual Report and Accounts. Retrieved from: https://assets.publishing.service.gov.uk/government/uploads/system/uploads/attachment_data/file/727762/oga-annual-report-accounts-2017-18-print.pdf [Accessed 20 December 2018].
44 Cameron, D., 2013, 11 August. We Cannot Afford to Miss Out on Shale Gas. *The Telegraph*. Retrieved from: www.telegraph.co.uk/news/politics/10236664/We-cannot-afford-to-miss-out-on-shale-gas.html [Accessed 20 November 2017].
45 Carrington, D., 2015, 25 January. George Osborne Urges Ministers to Fast-track Fracking Measures in Leaked Letter. *The Guardian*. Retrieved from: www.theguardian.com/environment/2015/jan/26/george-osborne-ministers-fast-track-fracking [Accessed 20 November 2017].
46 Department for Business, Energy and Industrial Strategy, 2017. New Measures to Back British Shale Gas Exploration. Retrieved from: www.gov.uk/government/news/new-measures-to-back-british-shale-gas-exploration [Accessed 20 December 2018].
47 Johnstone, P. et al., 2017. Policy Mixes for Incumbency: Exploring the Destructive Recreation of Renewable Energy, Shale Gas 'Fracking,' and Nuclear Power in the United Kingdom. *Energy Research and Social Science*, [online], 33 (147–162). Retrieved from: doi:10.1016/j.erss.2017.09.005 [Accessed 17 February 2019].
48 Ineos, 2017. Ineos Group Holdings Annual Report. Retrieved from: www.ineos.com/globalassets/investor-relations/public/annual-reports/2017-igh-sa-annual-report_final.pdf [Accessed 20 December 2018].
49 Food and Water Watch. The Awful Environmental Record of Ineos Disqualifies Fracking Ambitions. Retrieved from: www.foodandwaterwatch.org/news/awful-environmental-record-ineos-disqualifies-fracking-ambitions [Accessed 20 December 2018].
50 Sanderson, D., 2015, 9 April. Sturgeon held talks with Ineos chief as Government Called Halt to Fracking. *The Herald*. Retrieved from: www.heraldscotland.com/news/13209106.sturgeon-held-talks-with-ineos-chief-as-government-called-halt-to-fracking/ [Accessed 20 November 2017].
51 Graham, A., 2019, 4 February. Jim Ratcliffe: Government is using 'Slippery' Manoeuvres to Kill off British Fracking. *City AM*. Retrieved from: www.cityam.com/272673/jim-ratcliffe-government-using-slippery-manoeuvres-kill-off/ [Accessed 17 February 2019].

52 Newell, P. and Paterson, M., 1998. A Climate for Business: Global Warming, the State and Capital. *Review of International Political Economy*, [online], 5 (4). Retrieved from: www.jstor.org/stable/4177292?seq=1#page_scan_tab_contents [Accessed 17 February 2018].
53 Pirani, S., 2018. *Burning Up: A History of Fossil Fuel Consumption.* London: Verso Books.
54 Farand, C., 2018, 13 December. Inside the Tent: Big Polluters Work to Shape Paris Agreement Rules at the UN Climate Talks. *DeSmog Blog.* Retrieved from: www.desmogblog.com/2018/12/13/inside-tent-big-polluters-work-shape-paris-agreement-rules-un-climate-talks [Accessed 20 December 2018].
55 Heede, R. and Oreskes, N., 2016. Potential Emissions of CO2 and Methane from Proved Reserves of Fossil Fuels: An Alternative Analysis. *Global Environment Change*, [online], 36 (12–20). Retrieved from: www.sciencedirect.com/science/article/pii/S0959378015300637?via%3Dihub [Accessed 17 February 2018].
56 Newell, P. and Johnstone, P., 2018. The Political Economy of Incumbency. *In*: Skovgaard, J. and Asselt, H., eds. *The Politics of Fossil Fuel Subsidies and Their Reform.* Cambridge: Cambridge University Press, 72.
57 UN Environment., 2018. The Emissions Gap Report 2018: Executive Summary. UN Environment. Retrieved from: www.unenvironment.org/resources/emissions-gap-report-2018 [Accessed 7 January 2019].
58 Hope, M., 2018, 12 December. Countries that Blocked 'Welcoming' of Major Climate Science Report at UN Talks have Dozens of Delegates with Ties to Oil, Gas, and Mining. *DeSmog Blog.* Retrieved from: www.desmogblog.com/2018/12/12/countries-blocked-welcoming-major-climate-science-report-un-talks-have-dozens-delegates-ties-oil-gas-and-mining [Accessed 20 December 2018].
59 Mills, C. W., 1959. *The Power Elite.* New York: Oxford University Press, 10.
60 Johnstone, P. and Newell, P., 2018. Sustainability Transitions and the State. *Environmental Innovation and Sustainability Transitions*, [online], 27 (72–82). Retrieved from: doi:10.1016/j.eist.2017.10.006 [Accessed 17 February 2018].
61 Roberts, D., 2018, 8 September. California Just Adopted Its Boldest Energy Target Yet: 100% Clean Electricity. *Vox.* Retrieved from: www.vox.com/energy-and-environment/2018/8/31/17799094/california-100-percent-clean-energy-target-brown-de-leon [Accessed 17 February 2019].
62 Raworth, K., 2017. *Doughnut Economics: Seven Ways to Think Like a 21st Century Economist.* London: Random House Business Books; Trebeck, K. and Williams, J., 2019. *The Economics of Arrival: Ideas for a Grown-Up Economy.* London: Policy Press.
63 Newell, P. and Paterson, M., 1998. A Climate for Business: Global Warming, the State and Capital. *Review of International Political Economy*, [online], 5 (4). Retrieved from: www.jstor.org/stable/4177292?seq=1#page_scan_tab_contents [Accessed 17 February 2018].
64 Johnstone, P. and Newell, P., 2018. Sustainability Transitions and the State. *Environmental Innovation and Sustainability Transitions*, [online], 27 (72–82). Retrieved from: doi:10.1016/j.eist.2017.10.006 [Accessed 17 February 2018].

65 Newell, P., 2016. The Political Economy of Incumbency: Beyond Fossil-fueled Capitalism. Retrieved from: www.youtube.com/watch?v=nG-CPoMMU7QE [Accessed 20 December 2018].
66 Wells, P. and Nieuwenhuis, P., 2012. Transition Failure: Understanding Continuity in the Automotive Industry. *Technological Forecasting and Social Change*, [online], 79 (9). Retrieved from: doi:10.1016/j.techfore.2012.06.008 [Accessed 17 February 2019].
67 Newell, P. and Paterson, M., 1998. A Climate for Business: Global Warming, the State and Capital. *Review of International Political Economy*, [online], 5 (4). Retrieved from: www.jstor.org/stable/4177292?seq=1#page_scan_tab_contents [Accessed 17 February 2018].
68 Malm, A., 2016. *Fossil Capital: The Rise of Steam Power and the Roots of Global Warming*. London: Verso.
69 Geels, F. W., 2014. Regime Resistance against Low-Carbon Transitions: Introducing Politics and Power into the Multi-Level Perspective. *Theory, Culture and Society*, [online], 31 (5). doi:10.1177%2F0263276414531627 [Accessed 17 December 2018].
70 Roberts, C. et al., 2018. The Politics of Accelerating Low-carbon Transitions: Towards a New Research Agenda. *Energy Research and Social Science*, [online], 44 (304–311). Retrieved from: doi:10.1016/j.erss.2018.06.001 [Accessed 17 December 2018].
71 Geels, F. W., 2014. Regime Resistance against Low-Carbon Transitions: Introducing Politics and Power into the Multi-Level Perspective. *Theory, Culture and Society*, [online], 31 (5). doi:10.1177%2F0263276414531627 [Accessed 17 December 2018].
72 Reich, R. B., 2017. *Saving Capitalism: For the Many, Not the Few*. New York: Vintage, 176.
73 Newell, P. and Johnstone, P., 2018. The Political Economy of Incumbency. *In*: Skovgaard, J. and Asselt, H., eds. *The Politics of Fossil Fuel Subsidies and Their Reform*. Cambridge: Cambridge University Press, 72.
74 Johnstone, P. et al., 2017. Policy Mixes for Incumbency: Exploring the Destructive Recreation of Renewable Energy, Shale gas 'Fracking,' and Nuclear Power in the United Kingdom. *Energy Research and Social Science*, [online], 33 (147–162). Retrieved from: doi:10.1016/j.erss.2017.09.005 [Accessed 17 February 2019].
75 Heffron, R. et al., 2015. Resolving Society's Energy Trilemma through the Energy Justice Metric. *Energy Policy*, [online], 87 (168–176). Retrieved from: doi:10.1016/j.enpol.2015.08.033 [Accessed 17 February 2019].
76 Union of Concerned Scientists. The Clean Power Plan. Retrieved from: www.ucsusa.org/our-work/global-warming/reduce-emissions/what-is-the-clean-power-plan#bf-toc-0 [Accessed 20 December 2018].
77 Friends of the Earth, 2017. The Big Ask: How You Helped Make Climate Change History. Retrieved from: https://friendsoftheearth.uk/climate-change/big-ask-how-you-helped-make-climate-change-history [Accessed 20 December 2018].
78 Pirani, S. 2018. *Burning Up: A History of Fossil Fuel Consumption*. London: Verso Books, 23.48.
79 The Whitehouse, 2015. Remarks by the President in Announcing the Clean Power Plan. Retrieved from: https://obamawhitehouse.archives.gov/the-press-office/2015/08/03/remarks-president-announcing-clean-power-plan [Accessed 17 February 2019].

80 TUC, 2009. Congress 2009 Speech by Ed Miliband. Retrieved from: www.tuc.org.uk/research-analysis/reports/congress-2009-speech-ed-miliband [Accessed 17 February 2019].
81 The Whitehouse, 2019. Statement by President Trump on the Paris Climate Accord. Retrieved from: www.whitehouse.gov/briefings-statements/statement-president-trump-paris-climate-accord/ [Accessed 17 February 2019].
82 Union of Concerned Scientists. The Climate Deception Dossiers (2015). Retrieved from: www.ucsusa.org/global-warming/fight-misinformation/climate-deception-dossiers-fossil-fuel-industry-memos#.XFtGI1z7Q2w [Accessed 20 December 2018].
83 Mayer, J., 2016. *Dark Money: The Hidden History of the Billionaires Behind the Rise of the Radical Right*. New York: Anchor Books.
84 Sheldon Whitehouse, 2018. Whitehouse Delivers 200th 'Time to Wake Up' Climate Speech. Retrieved from: www.whitehouse.senate.gov/news/release/whitehouse-delivers-200th-time-to-wake-up-climate-speech [Accessed 17 February 2019].
85 Brulle, R. J., 2014. Institutionalizing Delay: Foundation Funding and the Creation of U.S. Climate Change Counter-movement Organizations. *Climatic Change*, [online], 122 (4). Retrieved from: https://link.springer.com/article/10.1007/s10584-013-1018-7 [Accessed 17 February 2019]; Skocpol, T., 2013. Naming the Problem. Harvard University. Retrieved from: https://scholars.org/sites/scholars/files/skocpol_captrade_report_january_2013_0.pdf [Accessed 20 December 2018].
86 Macalister, T. and Harvey, F., 2013, 19 July. George Osborne Unveils 'Most Generous Tax Breaks in World' for Fracking. *The Guardian*. Retrieved from: www.theguardian.com/politics/2013/jul/19/george-osborne-tax-break-fracking-shale-environment [Accessed 20 November 2017].
87 Department for Business, Energy and Industrial Strategy, 2017. New Measures to Back British Shale Gas Exploration. Retrieved from: www.gov.uk/government/news/new-measures-to-back-british-shale-gas-exploration [Accessed 20 December 2018].
88 Harabin, R., 2015, 17 December. UK Announces Cut in Solar Subsidies. *BBC*. Retrieved from: www.bbc.co.uk/news/business-35119173 [Accessed 20 November 2017].
89 Harabin, R., 2015, 9 November. Government Energy Policies 'Will Increase CO2 Emissions'. *BBC News*. Retrieved from: www.bbc.co.uk/news/science-environment-34767194 [Accessed 20 November 2017].
90 Johnstone, P. et al., 2017. Policy Mixes for Incumbency: Exploring the Destructive Recreation of Renewable Energy, Shale Gas 'Fracking,' and Nuclear Power in the United Kingdom. *Energy Research and Social Science*, [online], 33 (147–162). Retrieved from: doi:10.1016/j.erss.2017.09.005 [Accessed 17 February 2019].
91 Legislation.Gov.UK, 2018. The Feed-In-Tariffs (Closure). Retrieved from: www.legislation.gov.uk/uksi/2018/1380/article/6/made [Accessed 9 March 2019].
92 Johnstone, P. et al., 2017. Policy Mixes for Incumbency: Exploring the Destructive Recreation of Renewable Energy, Shale Gas 'Fracking,' and Nuclear Power in the United Kingdom. *Energy Research and Social Science*, [online], 33 (147–162). Retrieved from: doi: 10.1016/j.erss.2017.09.005 [Accessed 17 February 2019].

93 UK Government, 2017. The Clean Growth Strategy. Retrieved from: https://assets.publishing.service.gov.uk/government/uploads/system/uploads/attachment_data/file/700496/clean-growth-strategy-correction-april-2018.pdf [Accessed 9 March 2018].
94 Mayer, J., 2016. *Dark Money: The Hidden History of the Billionaires Behind the Rise of the Radical Right*. New York: Anchor Books.
95 Domhoff, G. W., 2017. Situating Who Rules America? Within Debates on Power. *In*: Domhoff, G. W., eds. *Studying the Power Elite*. Oakland: PM Press, 15.16.
96 Angus, I., 2016. *Facing the Anthropocene: Fossil Capitalism and the Crisis of the Earth System*. New York: Monthly Review Press.
97 Angus, I., 2016. *Facing the Anthropocene: Fossil Capitalism and the Crisis of the Earth System*. New York: Monthly Review Press.
98 Mills, C. W., 1959. *The Power Elite*. New York: Oxford University Press.
99 Phillips, K., 2003. *Wealth and Democracy: A Political History of the American Rich*. New York: Broadway Books.
100 Jaffe, J., 1993. *The Struggle for Market Power: Industrial Relations in the British Coal Industry, 1800–1840*. Cambridge: Cambridge University Press.
101 Church, R., 1986. *The History of the British Coal Industry: Victorian Pre-Eminence. Volume 3*. Oxford: Oxford University Press.
102 Hughes, M., 1963. *Lead, Land and Coal as Sources of Landlord Income in Northumberland between 1700 and 1850*. Newcastle: Newcastle University.
103 Milanovic, B., 2018, 28 January. There are Two Sides to Today's Global Income Inequality. *Globe and Mail*. Retrieved from: www.theglobeandmail.com/report-on-business/rob-commentary/the-two-sides-of-todays-global-income-inequality/article37676680/ [Accessed 20 November 2018]; Stiglitz, J., 2013. *The Price of Inequality: How Today's Divided Society Endangers Our Future*. New York: W. W. Norton & Company.
104 Atkinson, T., 2015. *Inequality: What Can Be Done?* Cambridge, MA: Harvard University Press.
105 Piketty, T., 2014. *Capital in the Twenty-First century*. Cambridge, MA: Harvard University Press.
106 Helm, T., 2017, 2 April. Osborne's Huge Tax Giveaway Starts for Rich – As the Poor are Hit. *The Guardian*. Retrieved from: www.theguardian.com/politics/2017/apr/01/huge-tax-giveaway-for-rich-as-poor-are-hit-george-osborne-tax-benefit-budget-changes [Accessed 20 November 2017].
107 HMRC, 2018. Corporation Tax Rates. Retrieved from: www.gov.uk/government/publications/rates-and-allowances-corporation-tax/rates-and-allowances-corporation-tax [Accessed 20 December 2018]; Institute for Fiscal Studies. Fiscal Facts: Tax and Benefits. Retrieved from: www.ifs.org.uk/tools_and_resources/fiscal_facts [Accessed 20 December 2018].
108 London Stock Exchange. Historic Market Capitalisation 2015. Retrieved from: www.londonstockexchange.com/statistics/historic/company-files/2015.zip [Accessed 20 December 2018].
109 Harringon, B., 2016. The Global 1 Percent Under Siege. *Roar Magazine*. Retrieved from: https://roarmag.org/wp-content/uploads/2017/04/ROAR_Issue_3_The_Rule_of_Finance.pdf [Accessed 20 November 2018].

110 Zucman, G., 2015. *The Hidden Wealth of Nations. The Scourge of Tax Havens.* Chicago: University of Chicago Press, 53.
111 Makdissi, P. and Yazback, M., 2015. On the Measurement of Plutonomy. *Social Choice and Welfare*, [online], 44 (4). Retrieved from: https://ideas.repec.org/a/spr/sochwe/v44y2015i4p703-717.html [Accessed 20 December 2018].
112 Sheldon Whitehouse, 2018. Whitehouse Delivers 200th 'Time to Wake Up' Climate Speech. Retrieved from: www.whitehouse.senate.gov/news/release/whitehouse-delivers-200th-time-to-wake-up-climate-speech [Accessed 17 February 2019].
113 Downey, L., 2015. *Inequality, Democracy and the Environment.* New York: New York University Press.
114 Boyce, J. K., 2018. How Economic Inequality Harms the Environment. *Scientific American.* Retrieved from: www.scientificamerican.com/article/how-economic-inequality-harms-the-environment/ [Accessed 20 December 2018].

5 Destabilising the polluter elite

The evidence presented above on the political power of the *polluter elite* demonstrates they have had a decisive effect in slowing down the transition away from fossil fuels. They have used their structural power over the state and their superior resources to counter policy measures that aimed to reduce greenhouse gas emissions. These findings confirm the growing recognition in debates on sustainability transitions of the importance of studying powerful incumbents who are blocking change because they fear they will lose out.[1]

This chapter will argue that to accelerate the transition the oil and gas *polluter elite* needs to be comprehensively destabilised. This could be done by ending fossil fuel subsidies, introducing a high carbon tax, phasing out fossil fuels and providing targeted support to renewable energy.

For policymakers in the US and UK to do this, they will need to feel sufficient public pressure to force them to challenge the *polluter elite.* This pressure would enable them to justify deliberately destabilising the commercially viable oil and gas sector in order to rapidly reduce emissions to avoid irreversible global warming.

5.1 The case for destabilisation

In 2019 the *polluter elite* still hold significant structural power over the state because the US and UK economies are still heavily reliant on oil and gas for economic growth. In the US, the fracking revolution has seen the country go from a net energy importer to become the largest oil and gas producer in the world. In his 2019 State of the Union speech, President Trump celebrated that, "The United States is now the number-one producer of oil and natural gas anywhere in the world. And now, for the first time in 65 years, we are a net exporter of energy".[2] In the UK, the Conservative government stresses dependence on oil and gas as an accepted state of affairs now and in the future. At the end of 2018, Minister of State for Energy and Clean Growth Claire Perry told the Scottish Affairs Committee that, "currently 80% of our homes are heated by gas and 65% use gas for cooking" and "every scenario we see for our cleaner future has some element of fossil fuels in the mix". She went on to say, "I guess the view is that if we are going to be using fossil fuels, we would like to use those that are generated from our domestic assets and that employ people in the United Kingdom".[3]

The *polluter elite* also have superior resources that they can deploy to counter attempts to accelerate the low-carbon transition. They and their companies have amassed vast amounts of wealth which they are willing to invest in donations to political parties, funding of fossil fuel lobbyists and in some cases to promote climate change denial. By paying lower tax (income and corporation) and using tax havens, the *polluter elite* and the companies they run are increasing the resources at their disposal. As the examples above highlight, it will take more than a set of well-intentioned policies to ensure fossil fuels are replaced by renewable energies. Even when the US has signed up to international treaties such as the Kyoto Protocol and the Paris Agreement the *polluter elite* have pressured for withdrawal. In the UK despite the Climate Change Act putting in place legally binding targets to reduce emissions, it did not deter or stop the Conservative government from continuing to give tax relief for fossil fuels and seeking to maximise oil and gas extraction.

The situation is circular. As long as fossil fuel companies run by the *polluter elite* are the largest and most profitable companies, they will continue to exert structural power over the state. As long as these companies have the green light from the government to continue extracting fossil fuels, they will continue to make a profit. As long as they continue to do this, they will use some of this profit to lobby the state to maintain fossil fuels in the energy mix alongside renewable energy (Figure 5.1).

The *polluter elite* are not going to go quietly. They have too much at stake and fear a fatal reduction in their personal net worth. They will never stop seeking political influence to deepen the dependency on fossil fuels. The incumbent regime needs to be "destabilised" so that renewable energies can replace fossil fuels (Turnheim and Geels, 2012).[4] If the oil and gas companies are not destabilised, then there will be no acceleration of the low-carbon transition that is needed to ensure emissions are reduced in line with what climate science demands. Whilst the transition towards a low-carbon economy has begun, mainly propelled by technological advances in solar and wind energy, it is not happening fast enough or on a large enough scale.

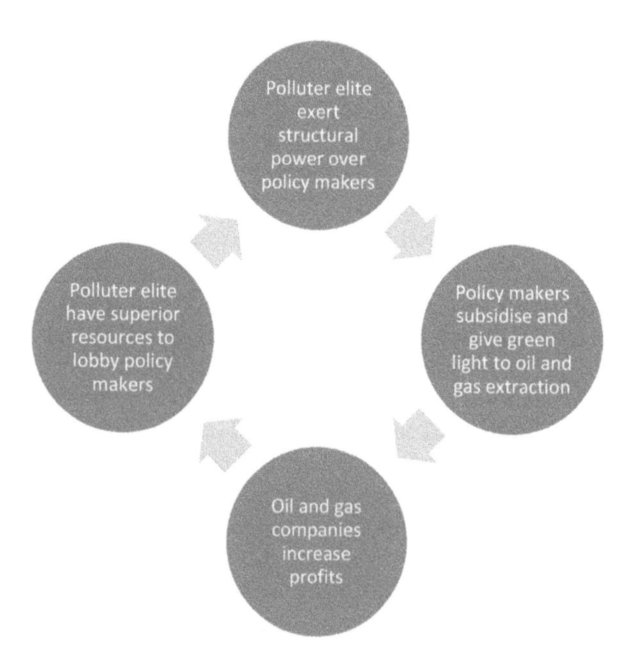

Figure 5.1 The structural power and superior resources of the polluter elite.
Source: Author.

Renewable energy use is growing in the US and the UK, but this is alongside the continued use of fossil fuels, in particular oil and gas.

Oil and gas need to be immediately destabilised to open up space for renewable energy to replace fossil fuels. Turnheim and Geels see destabilisation as crucial to accelerating the transition. Kivimaa and Kern build on this work to argue that sustainability transitions will only be achieved if policymakers push for a mix of measures which will both "destroy" the old regime as well as encourage typical policies for the "creation" of niche green technologies (Kivimaa and Kern, 2016).[5] They found that these policy mixes had been effective in the UK. The removal of subsidies for coal mining had made it easier for cleaner alternatives to emerge. Policies which on paper had destabilised the fossil fuel regime included the Climate Change Act with the creation of the independent and influential Climate Change Committee identified as a positive development. However, they also observed that the lack of further destabilisation policies was probably due to resistance. This fits with the above assessment of the political power of the *polluter elite*.

The IPCC concludes that greenhouse gas emissions need to fall rapidly by 2030 to avoid the worst impacts of climate change. There is not enough time to wait for alternatives to gradually replace fossil fuels and this is why active destabilisation is needed. Given this urgency how can the transition away from fossil fuels be accelerated? Perhaps there are lessons to learn from the experience of the destabilisation of the coal sector in both countries.

5.2 The destabilisation of coal

Destabilisation can happen in various ways. Incumbents can feel the effects of disruptive innovations, changing markets, new entrants, the loss of political, cultural and public legitimacy, or decisions made within the firm itself (e.g. inertia and slow response once destabilisation has begun).

5.2.1 The UK coal industry

In 1913 the British coal industry provided around 10% of jobs, employing just over one million people. It was also the basis of the fortunes of the historic *polluter elite* in the UK (see Box 2.1). Based on their study of the decline of the British coal industry during the twentieth century Turnheim and Geels found coal was destabilised by shrinking financial resources (direct cause of destabilisation) and the loss of political and public legitimacy (mediating factors in destabilisation).[6] The following factors were important:

The role of policymakers: Fluctuating support during the twentieth century highlights the influence of policymakers. Policymakers supported coal through subsidies and for it to be used in electricity generation. However, at various points over the twentieth century, policymakers reduced demand for coal by prioritising oil, gas and nuclear energy instead. During her time as Prime Minister, Margaret Thatcher had a huge impact on the coal industry. She broke the coal miners' strike and liberalised the energy market which was a driver in the "dash for gas" during the 1990s that saw newly privatised electricity distributors begin building new gas power stations.[7]

Competition from viable technological alternatives: Coal was undercut by competition from oil, gas and nuclear energy (in particular between 1913 and 1967). These alternatives destabilised coal by replacing it and opening up space for these alternatives to grow. When new alternatives become credible they enable policymakers to justify a shift towards them. By 2019 solar, wind and biomass have displaced coal production in the UK and were a contributing factor in coal-free hours (when no electricity was generated from coal) increasing from 2016. (This shift partly arose because government policies from 2010 onwards subsidised these alternative energies and supported them coming to fruition).

Competition from abroad: British coal found it hard to compete with other sources of coal. This led to declining exports in competition with cheaper imports. This pressure continued to be a factor in declining coal production after the industry was mainly in private hands from 1995.

Loss of public legitimacy: Environmentalists highlighted respiratory disease and deaths caused by coal smoke. In activities reminiscent of today's *polluter elite*, the coal industry countered this in the early twentieth century by lobbying policymakers to weaken regulations. They argued that smoke was not that damaging and stressed the importance of coal for job creation and economic growth (in a repetition of the energy security dimension of the energy trilemma covered in Chapter 4). While it took several decades for environmental issues to affect public opinion, once it did it was a powerful force. The 1956 Clean Air Act was passed following public outrage at 4,000 deaths in London related to smog in 1952. The Act reduced demand for coal, which was a form of economic destabilisation.

5.2.2 The US coal industry

In the US coal has been destabilised by similar factors. Production peaked in 2008 falling from 579.3 million metric tons to 348.5 million metric tons in 2016.[8] In between 2014 and 2016 several coal companies went bankrupt due to falling demand domestically and internationally. The overall trend has been one of rapid closure of coal mines and coal-fired power plants. However, there are significant differences between the Appalachian Basin which has seen the largest fall in production and the Powder River Basin which has increased production.[9]

The role of policymakers: The 1990 Clean Air Act of 1990 limited sulphur dioxide and nitrogen at power plants increasing costs as smokestack scrubbers had to be installed. After being introduced in 2015 President Obama's Clean Power Plan also weakened coal. President Trump has since reversed the measure and attempted to put in place tax breaks for the sector.

Competition from viable technological alternatives: Almost all coal consumed in the US is for electricity generation. Between 2008 and 2015 natural gas production jumped by 50% due to the fracking revolution.

Coal production fell by 28% over the same period.[10] Electricity from renewable sources has also grown. Renewables have become increasingly competitive as there are lower capital costs and there has been support through tax credits.

Competition from abroad: The global fall in coal prices between 2011 and 2015 hit US producers. Cheap imports have weakened domestic coal production. In 2015 coal imported from Colombia made up the vast majority of total imports.

5.3 How could oil and gas be destabilised? Deliberate destabilisation requires a perfect storm

As demonstrated above, powerful incumbents such as the coal industry can be destabilised. This highlights that it is possible. Coal went from being the foundation of the Industrial Revolution in the UK to being comprehensively destabilised. It is not inevitable that the US and UK economies continue to be dominated by oil and gas.

Turnheim and Geels argue coal in the UK was destabilised by a perfect storm of factors. This process is "multi-dimensional, involving interacting economic, technological, political, cultural, and business dimensions" (Turnheim and Geels, 2012).[11] Looking at another country, Germany, Kungl and Geels found that incumbents such as coal and nuclear energy were also destabilised over time by multiple interlinked pressures.[12] While these companies are sometimes able to deal with a single challenge, they can be overwhelmed by multiple problems occurring simultaneously.

Therefore, for the oil and gas companies to be comprehensively destabilised will require a similar perfect storm because as powerful incumbents they have already shown great capacity to resist change (see Chapter 4). I will now discuss each of the factors that have been identified as contributing to the destabilisation of coal, in relation to oil and gas.

5.3.1 Competition from viable technological alternatives

Out of all the factors this is the one that is most advanced. In 2019, the transition away from fossil fuel is becoming more and more possible. This is due to advances in technology that are making alternatives more efficient and cheaper. Solar and wind energy, battery storage, and electric vehicles are a credible alternative to oil and gas (and coal). For example, the costs of a solar panel have fallen by 99% since the mid-1970s making them competitive with fossil fuels in many countries.[13]

The price of batteries is also falling rapidly which is expected to make electric cars a lot more competitive.[14] New entrants in vehicle manufacturing, such as Tesla and Google, are already disrupting traditional vehicle manufacturers. The accumulation of technical knowledge, and as yet undeveloped technologies, suggest there is great potential to speed up the energy transition.[15] As there is already extensive literature and evidence of the capacity of renewable energy and clean technology I will now turn to the other key factors in more detail.

5.3.2 Loss of public legitimacy

If oil and gas companies lost public legitimacy this would contribute to pressure on policymakers to destabilise them. This could, in turn, increase the political will to accelerate the transition away from fossil fuels. However, the oil and gas companies have a proven track record of doing all they can to maintain their legitimacy and indeed these discursive strategies are a key component of their political influence.[16] In the past, they concentrated on distorting public understanding of climate science by funding climate change denial (some of the *polluter elite* such as the Koch brothers continue with this strategy). Now, these companies tend to focus on promoting gas as a cleaner fuel compared to coal[17] (in a repetition of history this is exactly what British companies did in the early twentieth century). These companies also focus their public image on their investments in biofuels, solar and wind energy, which is an indicator of them accepting disruption by renewables up to a certain point (proving there is credible competition from viable technological alternatives).

While oil and gas companies have manoeuvred to maintain legitimacy, it will not necessarily always be this way. Roberts studied the collapse of the American railroad from the 1920s to 1940s as it was replaced by motorised road transport. He argues that negative storylines were a contributing factor in this shift. Powerful incumbents are vulnerable to these negative storylines gaining traction and becoming a focal point for anti-regime coalitions, all of which can then open up space for alternatives. This could happen with fossil fuels as worries about energy bills combine with concerns about air pollution from coal-fired power plants and from fracking for shale gas. According to Roberts existing incumbents could be seen as "dangerous, ecocidal, and out-of-touch monopolies" (Roberts, 2017).[18] In terms of what has already happened, the past few years have seen civil society responses, such as campaigns for fossil fuel divestment and protests to block fossil fuel infrastructure, that explicitly seek to question the social licence to operate of oil and gas companies. Questioning the legitimacy of investments in polluting activities as a form of wealth creation could be another component of destabilising the legitimacy of the *polluter elite* (see Chapter 3).

5.3.3 The role of policymakers

If policymakers decided to destabilise oil and gas companies, they would have to implement effective measures to do so. How could this happen in practice? The US and UK government could end fossil fuel subsidies worth billions of dollars a year, place a carbon tax on the operations of oil and gas multinationals, and ultimately phase out oil and gas extraction.

5.3.3.1 Ending subsidies for fossil fuels and applying a carbon tax

The US and UK governments could remove subsidies for fossil fuels and also introduce a carbon tax. Both of these measures would, in theory, see fossil fuels compete in a "fairer" way with renewable energy. As renewable energy (solar and wind) is already competitive, or getting there, in the US and the UK, the logic is this would further tilt investment towards renewable energy.

The US and UK government could end all forms of subsidies and tax breaks for fossil fuels which are worth billions of dollars. These subsidies could instead be used to promote renewable energy and an overall shift away from fossil fuels. Subsidies for the solar industry in the UK was one factor that saw the sector grow rapidly until this support was cut by the Conservative government from 2015 onwards which led to falling investment.

Removing fossil fuel subsidies is a prerequisite if the *polluter elite* are to be destabilised but it should not be seen as one solution in isolation. As Newell and Johnstone reflect: "We might prune the branches and dead leaves with fossil fuel subsidy reform, but the trunk of the tree (or the fossil fuel economy) could remain sturdy, with roots that spread and branches that grow back in different directions" (Newell and Johnstone, 2018).[19] What other ways could the *polluter elite* be destabilised?

The goal of carbon pricing is to incentivise a long-term reduction in greenhouse gas emissions by making it more expensive to pollute (some scholars see this as a more cost-effective measure because actors could choose the cheapest way to reduce their emissions). In effect, those actors that currently pollute for free would be forced to pay for the pollution they impose on others. In theory it would make the operations of the companies that are run by the *polluter elite* more expensive and therefore reduce their profits. This could be done by capping emissions and shrinking allowances to pollute (cap-and-permit system) or by taxing pollution (carbon tax). There are ongoing debates about how to calculate the "correct" price with some advocating an approach based on optimal efficiency using economic models ("social cost of carbon" approach which often uses discount rates to translate future damage into present values), and others that the price be guided by the costs of meeting emissions targets as set out in the Paris Agreement.[20]

Carbon pricing could be applied at the moment in the supply chain when fossil fuels enter the economy through mines, pipelines and ports. In the US this has already been explored for around 2000 collection points.[21] In the UK there is already a carbon tax in place (its official name is carbon price support) that was introduced in 2013. The main impact has been to make gas-fired power stations cheaper to run because they are more efficient and release less carbon dioxide emissions than coal-fired power stations.

However, there are key considerations if the US or UK government were to apply either of these measures because in addition to affecting the *polluter elite* they could, depending on how they are designed, also affect households. Experience shows that if a government does not engage with the social repercussions of these types of measures then they can be incredibly difficult to implement and may backfire. The principal issue here is how these measures will affect the majority of the population, in particular, the poorest because it is often lower-income households who spend a greater percentage of their income on fuel. This has led for calls to share the revenue collected with citizens (known as cap-and-dividend) to compensate them for the higher costs of goods and services.

If the above measures do not factor in issues of equity they can result in sections of the population feeling as if they have been "pushed unreasonably and too far" to the extent they can end up opposing the transition and perpetuating the fossil fuel status quo.[22] For example, the Labour government was forced to abandon attempts to remove fuel duty in 2001 after lengthy protests, mainly by lorry drivers. Fuel duty has been frozen for many years and politicians do not want to anger motorists. In the US President Obama said before the 2008 and 2012 elections that he would keep petrol prices down. This dynamic is played out in many countries around the world and that is why policymakers are wary of raising fuel prices because they know it will likely lead to protests. One of the clearest examples of this was the sustained protests against fuel tax rises (as well as reduction of taxes on the wealthy) in late 2018 which forced French President Emmanuel Macron to revoke the measure.[23]

5.3.3.2 Phasing out of fossil fuels

Even if the oil and gas companies were destabilised by the removal of fossil fuel subsidies and carbon tax they would still seek to extract fossil fuels. History shows that at the end of the cycle of destabilisation, that is, when a company knows it is in decline and cannot prevent this, it will still try to avoid fully ending its operations and instead "milk" its assets as much as possible.[24] A phaseout of fossil fuels should aim to avoid the "green paradox" where companies increase extraction of fossil fuels in anticipation of expected policies to reduce emissions. If this were to happen it would mean that greenhouse gas emissions do not fall fast enough to keep within 1.5°C of global warming.

This is why the *polluter elite* have advocated for neo-liberal market mechanisms because they understand that one of the alternatives is a phaseout of fossil fuels. They prefer to take a financial hit via a carbon tax and to be able to continue their operations. For example, several oil and gas companies have publicly embraced the policy of a carbon tax because they do not see it as fully disrupting their business model. They continue to make public statements stating they see a role for fossil fuels in the future. In 2015 Ben van Beurden, CEO of Shell, said:

> I think we will get to the point where we have zero emissions by the end of the century, definitely, I am a firm believer in that, but even then some of the hydrocarbons that we will use and the emissions that will come from it will simply be mitigated rather than not produced.[25]

The main reason to phase out fossil fuels would be to contribute to meeting climate targets to avoid irreversible global warming. The IPCC report published in 2018 concluded emissions needed to fall dramatically by 2030. The only guarantee this would happen is mandated reductions in the use of fossil fuels alongside other complementary policies such as carbon pricing.[26] A phaseout would also be one of the most effective ways to permanently cut off the source of economic and political power of the *polluter elite*.

Unless the state regulates to phase out fossil fuels completely, they will continue to be extracted as long as they are profitable. This could mean that, as is currently the case, fossil fuels are used alongside growing renewable energy. This would be a continuation of what we currently have where the brown economy continues alongside the incipient, but growing, green economy. In this scenario the *polluter elite* will continue to exert political influence and fossil fuels will continue to be extracted and burnt. The state has a key role to play in destabilising the *polluter elite*. It is the only actor with the authority and capacity to put in place the measures needed to deliberately accelerate the transition on the large scale that is needed by phasing out fossil fuels. The state has the ability to exert authority over market actors and directly intervene in markets in ways which would destabilise the *polluter elite*.[27] Phasing out may appear an extreme measure, but it is already under discussion. Whilst too far in the future to have much impact now the G7 did announce at the 2015 summit that it would phase out fossil fuels by 2100 (although clearly, the election of President Trump undermines this commitment on the part of the US).

How could oil and gas (and coal) be phased out in practice? The extraction of oil, gas and coal could be banned. An announcement could be made to close gas-fired power stations. A deadline could be given for a shift to electric vehicles. In the UK the government has already announced the banning of petrol and diesel vehicles sales from 2040, and is under pressure to bring this forward to 2032.[28]

More recently, there are signs some policymakers in the US are considering backing a form of phasing out of fossil fuels via a Green New Deal.[29] This is a wide-ranging proposal for the state to fund measures to stimulate the economy via investments, in areas such as renewable energy, to create new green jobs, infrastructure, products and services. Whilst the specifics of the Green New Deal are still being defined the version of the Deal presented by Congresswoman Alexandria Ocasio-Cortez calls for the decarbonisation of the US economy in ten years, for example by switching to 100% renewable electricity.[30]

If policymakers were to phase out oil and gas they would need to put in place measures to support the people who rely on these companies for employment (directly and indirectly). There are lessons to be learnt from how this has and has not happened with coal workers. In the UK successive governments have given welfare benefits to former coal miners. However, there is still high unemployment in these areas and where there are jobs they have often been low-paid and insecure.[31] President Obama did put in place the Power + Plan which targeted job creation including in solar panel installation and energy efficiency.[32] If oil and gas were phased out there would need to be solid plans to deal with unemployment otherwise these policies could lead to social conflict and reduced public support.[33] These themes are central to the concept of a "just transition" which combines emissions reductions with social equity to ensure the costs are not disproportionately felt by poorer and more vulnerable groups.[34] In this context it is interesting to see that one of the explicit goals of the Green New Deal is that it would guarantee decent jobs. This is probably one of the reasons behind its popularity with a wide range of voters.[35] There is significant potential for job creation in renewable energy. In 2018 there were around 242,000 full-time jobs and 93,000 part-time jobs in solar energy firms in the US.[36]

5.3.3.3 Support green innovation

As mentioned above the other side of the coin to destabilisation is that policymakers could actively support renewable energies to replace fossil fuels. How could they do this? Policymakers could accelerate the transition by coordinating the design for a new energy system. Solar and wind energy are increasingly cost-competitive with fossil fuels and have the potential to be deployed on a large-scale. This could see a tipping point in 2035 when renewable energies account for 50% of newly installed generation capacity or capture 20% of the market share.[37] However, for solar and wind to be scaled up means a change in the overall energy system. Policymakers would need to actively coordinate to update the flexibility of the centralised energy system so as to maximise the use of intermittent renewables. This is not simply a change in energy source, it is a change in how and where energy is produced and consumed overall. For example, there could be more decentralised local energy production and consumption. This will require enormous levels of coordination to plan sites of solar and wind energy generation (primary renewable supply), upgrade national grids to store and distribute renewable energy efficiently (distribution of secondary energy), and to organise the shift to electric cars.[38]

Storage is absolutely fundamental to the deployment of a new energy system. It is an area where innovation is crucial to increase the capacity to store intermittent renewables over longer periods of time.[39] This illustrates the key role policymakers can play in promoting green innovation by funding research and development for a post-fossil fuel economy. It was public funding for the US Department of Defense agencies on advanced research projects (DARPA and ARPA-E) that led to advances in technology on fracking, nuclear, solar and battery storage.[40] There is an assumption that state policy should be used to make an even playing field (i.e. not to pick winners) but in the area of green innovation, it is essential that the state directly supports it.[41]

Indeed, historically this is what has happened. It has been the government, alongside society and business, that has shaped and steered the major technological revolutions of previous transitions which have then enabled new interrelated infrastructure and production systems for new goods and services.[42]

5.4 The destabilisation of oil and gas will need to be different from that of coal

While the destabilisation of coal provides many useful lessons, this experience does not directly translate to the current situation with oil and gas. Firstly, there is an issue of speed. It took several decades for some of the key factors to build up into a perfect storm to destabilise coal. Due to the urgency to immediately reduce emissions to avoid irreversible climate change there is not enough time to wait for the oil and gas companies to be destabilised gradually. The transition needs to be deliberately accelerated.[43] The shift to renewable energy will need to happen a lot faster than previous energy transitions. The rise in the use of oil, gas and coal in the US each took between 70 to 100 years.[44]

Secondly, when policymakers in the UK deprioritised coal from the 1980s onwards it had already been economically weakened by a combination of factors such as cheaper competition from abroad and competition with credible technological alternatives such as gas. The oil and gas industries are not in the same position as coal was. They are commercially viable industries with huge profits which is why they continue to exert such strong structural power over the state and have superior resources to lobby policymakers. Oil and gas have not been sufficiently economically destabilised by renewable energy (or other sources such as nuclear) as was the case when gas filled the gap left by coal.

Finally, coal has been destabilised in both countries, but this does not mean it has disappeared altogether. In the UK the Conservative government has announced the phaseout of coal-fired power stations by 2025. However, coal is still used to generate electricity and coal extraction continues. The Banks Group began operations at the Pont Valley open-cast mine in mid-2018 and is seeking to open a new open-cast mine in Druridge Bay in Northumberland.[45] In the US coal is still extracted with production of 348.5 million metric tons in 2016. President Trump attempted to put in place subsidies for the sector which could have supported higher extraction. However, courts ruled against this. Therefore, despite being destabilised economically coal is still in use in both countries. It might be more accurate to refer to this as an economic decline.[46] This matters for the climate because coal is the most polluting energy source. It is also an indicator of how difficult it is to completely phase out fossil fuels. The same is the case with oil and gas. Even if they are somehow destabilised in the coming years they will seek to continue operating as long as they can. The problem is that this delays the full transition to a low-carbon economy and will see increasing greenhouse gas emissions that add to the already accumulated emissions in the atmosphere. As the IPCC report published in 2018 indicated: "Limiting global warming requires limiting the total cumulative global anthropogenic emissions of CO2 since the preindustrial period, that is, staying within a total carbon budget".[47]

In summary, what the previous Chapters show is that to date the *polluter elite* have adapted to the threat of renewable energy, and have successfully countered moments when they have lost political and public legitimacy to maintain the supremacy of fossil fuels. The lesson here is that unless the oil and gas companies (and the *polluter elite* who run them) feel economic pressure, they will continue business as usual. If the wealth of the *polluter elite*, and of the companies they run, is not reduced then they will be able to continue to use their significant political influence to block attempts to accelerate the low-carbon transition. The US and UK government need to deliberately destabilise the oil and gas industry in order to accelerate the low-carbon transition as part of global efforts to prevent a damaging 1.5-degree increase of average surface temperature.

5.4.1 Prospects for public pressure

The experience of past and ongoing destabilisation of coal in the US and the UK, as well as other countries such as Germany,[48] show that it is a process which generates conflict. The ability of the *polluter elite* to effectively slow down attempts to accelerate the transition show that progress is fragile and not linear. The policies and technology to reduce emissions are well developed. Disrupting the deeply embedded oil and gas sector is ultimately a political decision.[49] Policymakers will only proactively destabilise the *polluter elite* if they feel the pressure to do so.

To successfully push policymakers to destabilise the *polluter elite*, public pressure will need to be:

- **Sufficient**: For any of the destabilisation measures suggested earlier to be implemented would require overwhelming public pressure to counter the structural power and superior resources of the *polluter elite*. This could be the crucial factor which could

tip the balance to force policymakers to enact policies for destabilisation and be able to justify them to a wider public. The real test will be if there is sufficient public pressure to force the Republican and Conservative parties to destabilise the *polluter elite* because both these parties have approved policies slowing down the transition.
- **Sustained over the long-term**: Measures to reduce emissions which would destabilise the elite fall under the category of public interest policies. When it comes to the implementation phase of this legislation it is likely to be opposed by special interests, such as the *polluter elite*, because they will lose out. The problem is that the costs can be more visible than the benefits to the wider public. Climate change is seen as abstract which makes it more difficult for coalitions that want to accelerate the transition to defend these policies and easier for coalitions that want to undermine them to appear. Unless the power of incumbents is reduced then the policy change alone will not be enough to "upset existing power monopolies" and is at risk of being undermined because it is difficult to dislodge political opponents completely (Lockwood, 2013).[50]

What form could successful public pressure take? To further understand how sustainability transitions can be accelerated, Roberts et al recommend a future research agenda exploring how coalitions that wish to accelerate the transition could overcome incumbent coalitions.[51] Such progressive coalitions already exist and will undoubtedly play a crucial role. A range of civil society and business groups supported President Obama's Clean Power Plan and ratification of the Paris Agreement in 2015. When President Trump announced the withdrawal from the Paris Agreement coalitions of civil society, business and sub-national government (22 US states to date) committed to continue to meet emissions reduction goals.[52] This sub-national level government activity to reduce emissions is significant because it has potential to disrupt carbon lock-in further afield as it is linked to global energy and transport supply chains.[53] In the UK coalitions of civil society and business supported the Climate Change Act, the introduction of the FiT and the Paris Agreement.

Civil society organisations play a key role in existing coalitions. It can be common to see the low-carbon transition as a technical challenge to be introduced to the population but actually civil society, and in particular environmentalists, have done much, as Smith comments, to "propel governments and business into contemplating low carbon futures" in the first place (Smith, 2012).[54] It is social movements who can influence shifts in values by "creating new semiotic maps of the possible and desirable" (Köhler et al, 2019).[55] Therefore, it is important to understand their potential in destabilising incumbents but to date this is an under-researched area. This is a notable gap because social action by civil society has played a crucial role in shaping systems of energy production and consumption, for example, slowing the expansion of nuclear energy in the 1970s and 1980s.

BOX 5.1 THE CHALLENGES OF INCREASING PRESSURE

Despite the examples of public pressure and protests the majority of the US and UK population have not become actively involved in pressure against fossil fuels. Why is this? The obvious answer is that as pretty much the entire economy is based on fossil fuels the consumption habits of the population are highly skewed towards dependence on fossil fuels. Why would the majority of people in the US and the UK protest against fossil fuels when they depend on them for their electricity, transport and food? (indeed, as the above examples of protests over fuel price increases show there can be protests to defend the fossil fuel status quo). This is known as behavioural lock-in. Citizens' habits and norms have built up around the use of goods and services that require energy which to date has mainly been from fossil fuels.

One feature of behavioural lock-in is that in the popular imagination there is a strong link between oil and individual freedom. This energy enables people to control their own space such as driving wherever they want to (what Huber refers to as privatised geographies[56]). These are the everyday practices which make up people's lives and this is why it has been so difficult to move away from them despite the clear damage to the environment. Reliance on cheap natural resources and environmental degradation are the bases for the imperial mode of living.[57] High-carbon lifestyles have become a fundamental part of people's identity, for example they see themselves as highly mobile people. The "addiction" to oil is real in that it has become embedded in almost everything (which is why environmental movements must be able to offer alternatives to cheap energy), but this does not mean it is inevitable.

There are signs that public pressure on policymakers to address the environmental crisis is building. In recent years civil society has organised direct action to block fossil fuel operations[58] which has increased costs for the *polluter elite*. In the US, there were lengthy protests against the Keystone XL oil pipeline and at Standing Rock against the Dakota Access oil pipelines. These protests played an important role in convincing Obama to block both projects after several years of indecision.[59] This direct action does not always have a long-term impact. In an example of the continued power of the *polluter elite* President Trump approved both pipelines in 2017. However, there are other examples of protestors successfully blocking construction of a large coal export terminal in Washington State, temporarily blocking another coal terminal in Oakland, and in Montana after 30 years the Blackfeet Nation successfully had energy leases on 23,000 acres cancelled. What these examples show is that the balance of power can change when public pressure increases.[60]

In the UK protestors have blocked the operations of fracking companies which has led to lengthy delays and costs for these companies. Fracking has been slowed down but was still allowed to begin in Lancashire. It has been halted recently because of causing tremors over the seismic level of 0.5 magnitude. The government confirmed in February 2019 it had no intention to change this regulation, which fracking companies such as Cuadrilla had complained hindered their operations. Polls conducted at the end of 2018 show lower and lower support for fracking. Cultural legitimacy is one key factor which explains why there has been greater resistance to fracking in the UK compared to the US.[61] The political power of the *polluter elite* has been a contributing factor to the Conservative government consistently backing this nascent fracking industry. For example, it has repeatedly sought to relax planning rules to allow it to go ahead despite opposition from local councils and communities. The future of fracking is uncertain but what is clear is that this public pressure has slowed it down.

Downey argues that because the elite control the world's most important national and international decision-making bodies, what is needed are "social and environmental activists to adopt highly confrontational political tactics, such as mass demonstrations and other direct actions, that go outside normal political channels and place extreme pressure on elites and the organizations, institutions, and networks they control" (Downey, 2015).[62]

Other examples of civil society pressure are that young people are increasingly making their voice heard. As the generation that will live with the consequences of climate change, this makes sense. Students in the US and the UK have called on their universities to divest their endowment funds from fossil fuels. They have had considerable success with dozens of universities divesting their funds.[63] Schoolchildren in the US and the UK are starting to go on "strike" as part of global protests to demand that the government take action. An estimated 15,000 schoolchildren across the UK joined the first such strike on 15 February 2019.[64] The two MPs who convened a Parliamentary debate on climate change a few weeks later said they did so because of this climate strike. One of the demands of the schoolchildren is that the government declares a climate emergency, something 27 local authorities in the UK had done by February 2019 through non-binding motions, as well as the London Assembly.[65] If more policymakers were to declare a climate emergency this could potentially give them the legitimacy to justify the destabilisation of oil and gas companies.

In the US the Sunrise youth movement has pressured for the adoption of a Green New Deal. Hundreds of young people occupied the office of the Democrat Nancy Pelosi, speaker of the House of Representatives, in November 2018. The Green New Deal is gaining increasing traction in the US and could be a convergence point for coalitions of diverse groups to accelerate the transition. Whether this Green New Deal is made a reality will be one very visible test of whether the *polluter elite* have been destabilised by overwhelming public pressure.

5.5 Conclusion

The transition to a low-carbon economy in the US and the UK needs to be rapidly accelerated as part of worldwide efforts to avoid a 1.5-degree increase in global average surface temperature. As the experience of the coal sector demonstrates it is the policymakers who hold the key to destabilising the oil and gas *polluter elite*. They could end fossil fuel subsidies, tax pollution and ultimately cut off the source of the *polluter elite*'s wealth by phasing out fossil fuels. If these measures are not implemented the *polluter elite* will continue to use their wealth to lobby the state to maintain fossil fuels in the energy mix alongside renewable energy. It is in this context that civil society has a pivotal role in influencing whether policymakers will implement policies to undermine the *polluter elite*.

Notes

1 Roberts, C. et al., 2018. The politics of accelerating low-carbon transitions: towards a new research agenda. *Energy Research and Social Science*, [online], 44 (304–311). Retrieved from: doi:10.1016/j.erss.2018.06.001 [Accessed 17 December 2018].

2 The Whitehouse, 2019. Remarks by President Trump in State of the Union Address. Retrieved from: www.whitehouse.gov/briefings-statements/remarks-president-trump-state-union-address-2/ [Accessed 17 February 2019].

3 Parliament, 2018. The Future of the Oil and Gas Industry. *Parliament*. Retrieved from: https://publications.parliament.uk/pa/cm201719/cmselect/cmscotaf/996/996.pdf [Accessed 17 February 2019].
4 Turnheim, B. and Geels, F. W., 2012. Regime Destabilisation as the Flipside of Energy Transitions: Lessons from the History of the British Coal Industry (1913–1997). *Energy Policy*, [online], 50 (35–49). doi:10.1016/j.enpol.2012.04.060 [Accessed 17 December 2018].
5 Kivimaa, P. and Kern, F., 2016. Creative Destruction or Mere Niche Support? Innovation Policy Mixes for Sustainability Transitions. *Research Policy*, [online], 45 (1). doi:10.1016/j.respol.2015.09.008 [Accessed 17 December 2018].
6 Turnheim, B. and Geels. F. W., 2012. Regime Destabilisation as the Flipside of Energy Transitions: Lessons from the History of the British Coal Industry (1913–1997). *Energy Policy*, [online], 50 (35–49). doi:10.1016/j.enpol.2012.04.060 [Accessed 17 December 2018].
7 Fothergill, S., 2017. Coal Transition in the United Kingdom. *Climate Strategies*. Retrieved from: www.iddri.org/sites/default/files/PDF/Publications/Catalogue%20Iddri/Rapport/201706-iddri-climatestrategies-coal_uk.pdf [Accessed 7 January 2019].
8 International Energy Agency. Energy Atlas. Retrieved from: http://energyatlas.iea.org/#!/tellmap/2020991907 [Accessed 17 February 2019].
9 Kok, I., 2017. Coal Transition in the United States. *Climate Strategies*. Retrieved from: www.iddri.org/sites/default/files/PDF/Publications/Catalogue%20Iddri/Rapport/201706-iddri-climatestrategies-coal_us.pdf [Accessed 7 January 2019].
10 Culver, W. and Mingguo, H., 2016. Coal's Decline: Driven by Policy or Technology? *The Electricity Journal*, [online], 29 (7). doi:10.1016/j.tej.2016.08.008 [Accessed 17 December 2018].
11 Turnheim, B. and Geels, F. W., 2012. Regime Destabilisation as the Flipside of Energy Transitions: Lessons from the History of the British Coal Industry (1913–1997). *Energy Policy*, [online], 50 (35–49). doi:10.1016/j.enpol.2012.04.060 [Accessed 17 December 2018].
12 Kungl, G. and Geels, F. W., 2012. Sequence and Alignment of External Pressures in Industry Destabilisation: Understanding the Downfall of Incumbent Utilities in the German Energy Transition (1998–2015). *Environmental Innovation and Society Transitions*, [online], 26 (78–100). doi:10.1016/j.eist.2017.05.003 [Accessed 17 December 2018].
13 Carbon Tracker, 2017. Expect the Unexpected: The Disruptive Power of Low-carbon Technology. *Carbon Tracker*. Retrieved from: www.carbontracker.org/reports/expect-the-unexpected-the-disruptive-power-of-low-carbon-technology/ [Accessed 20 December 2018].
14 Bond, K. et al., 2019. The Political Tipping Point: Why the Politics of Energy Will Follow the Economics. *Carbon Tracker*. Retrieved from: www.carbontracker.org/wp-content/uploads/2019/01/The-Political-Tipping-Point-1.pdf [Accessed 20 December 2018]; Bloomberg New Energy Finance, 2018. Tumbling Costs for Wind, Solar, Batteries Are Squeezing Fossil Fuels. *BNEF*. Retrieved from: https://about.bnef.com/blog/tumbling-costs-wind-solar-batteries-squeezing-fossil-fuels/ [Accessed 20 December 2018].
15 Sovacool, B. K., 2016. How Long Will It Take? Conceptualizing the Temporal Dynamics of Energy Transitions. *Energy Research and Social Science*, [online], 13 (202–215). doi:10.1016/j.erss.2015.12.020 [Accessed 17 December 2018].

16 Geels, F. W., 2014. Regime Resistance against Low-Carbon Transitions: Introducing Politics and Power into the Multi-Level Perspective. *Theory, Culture and Society*, [online], 31 (5). doi:10.1177%2F0263276414531627 [Accessed 17 December 2018].
17 Johnstone, P. et al., 2017. Policy Mixes for Incumbency: Exploring the Destructive Recreation of Renewable Energy, Shale Gas 'Fracking,' and Nuclear Power in the United Kingdom. *Energy Research and Social Science*, [online], 33 (147–162). Retrieved from: doi:10.1016/j.erss.2017.09.005 [Accessed 17 February 2019].
18 Roberts, J. C. D., 2017. Discursive Destabilisation of Socio-technical Regimes: Negative Storylines and the Discursive Vulnerability of Historical American Railroads. *Energy Research and Social Science*, [online], 31 (86–99). doi:10.1016/j.erss.2017.05.031 [Accessed 17 December 2018].
19 Newell, P. and Johnstone, P., 2018. The Political Economy of Incumbency. *In*: Skovgaard, J. and Asselt, H., eds. *The Politics of Fossil Fuel Subsidies and Their Reform*. Cambridge: Cambridge University Press, 67.
20 Boyce, J. K., 2018. Carbon Pricing: Effectiveness and Equity. *Ecological Economics*, [online], 150 (52–61). Retrieved from: doi:10.1016/j.ecolecon.2018.03.030 [Accessed 17 December 2018].
21 Boyce, J. K., 2018. Carbon Pricing: Effectiveness and Equity. *Ecological Economics*, [online], 150 (52–61). doi:10.1016/j.ecolecon.2018.03.030 [Accessed 17 December 2018].
22 Smith, A., 2012. Civil Society in Energy Transitions. *In*: Verbong, G. and Loorbach, D., eds. *Governing the Energy Transition: Reality, Illusion, or Necessity*. New York: Routledge, 157.
23 Malm, A., 2018. *A Lesson in How Not to Mitigate Climate Change*. Verso. Retrieved from: www.versobooks.com/blogs/4156-a-lesson-in-how-not-to-mitigate-climate-change [Accessed 17 February 2019].
24 Turnheim, B. and Geels, F. W., 2012. Regime Destabilisation as the Flipside of Energy Transitions: Lessons from the History of the British Coal Industry (1913–1997). *Energy Policy*, [online], 50 (35–49). doi:10.1016/j.enpol.2012.04.060 [Accessed 17 December 2018].
25 Mathiesen, K., 2015, 8 June. G7 Fossil Fuel Pledge is a Diplomatic Coup for Germany's 'Climate Chancellor'. *The Guardian*. Retrieved from: www.theguardian.com/environment/2015/jun/08/g7-fossil-fuel-pledge-is-a-diplomatic-coup-for-germanys-climate-chancellor [Accessed 20 November 2017].
26 Boyce, J. K., 2018. Carbon Pricing: Effectiveness and Equity. *Ecological Economics*, [online], 150 (52–61). Retrieved from: doi:10.1016/j.ecolecon.2018.03.030 [Accessed 17 December 2018].
27 Johnstone, P. and Newell, P., 2018. Sustainability Transitions and the State. *Environmental Innovation and Sustainability Transitions*, [online], 27 (72–82). Retrieved from: doi:10.1016/j.eist.2017.10.006 [Accessed 17 February 2018].
28 BBC, 19 October. How Will the Diesel and Petrol Ban Work? *BBC News*. Retrieved from: www.bbc.co.uk/news/uk-40726868 [Accessed 20 November 2017].
29 Roberts, D., 2019, 23 February. The Green New Deal, Explained. *Vox*. Retrieved from: www.vox.com/energy-and-environment/2018/12/21/18144138/green-new-deal-alexandria-ocasio-cortez [Accessed 9 March 2019].
30 Ocasio-Cortez, A., 2019. Draft Text for Proposed Addendum to House Rules for 116th Congress of the United States: Final Select Committee for a Green New Deal. Retrieved from: https://docs.google.com/

document/d/1jxUzp9SZ6-VB-4wSm8sselVMsqWZrSrYpYC9slHKLzo/edit#heading=h.z7x8pz4dydey [Accessed 9 March 2019].
31 Fothergill, S., 2017. Coal Transition in the United Kingdom. *Climate Strategies*. Retrieved from: www.iddri.org/sites/default/files/PDF/Publications/Catalogue%20Iddri/Rapport/201706-iddri-climatestrategies-coal_uk.pdf [Accessed 7 January 2019].
32 Kok, I., 2017. Coal Transition in the United States. *Climate Strategies*. Retrieved from: www.iddri.org/sites/default/files/PDF/Publications/Catalogue%20Iddri/Rapport/201706-iddri-climatestrategies-coal_us.pdf [Accessed 7 January 2019].
33 Kungl, G. and Geels, F. W., 2012. Sequence and Alignment of External Pressures in Industry Destabilisation: Understanding the Downfall of Incumbent Utilities in the German Energy Transition (1998–2015). *Environmental Innovation and Society Transitions*, [online], 26 (78–100). doi:10.1016/j.eist.2017.05.003 [Accessed 17 December 2018].
34 Heffron, R. and McCauley, D., 2018. What is the Just Transition? *Geoforum*, [online], 88 (74–77). doi:10.1016/j.geoforum.2017.11.016 [Accessed 17 December 2018]; Climate Justice Alliance. A Green New Deal Must be Rooted in a Just Transition. Retrieved from: https://climatejusticealliance.org/green-new-deal-must-rooted-just-transition-workers-communities-impacted-climate-change/ [Accessed 17 February 2019].
35 Relman, E., 2019, 14 February. More than 80% of Americans support almost all of the key ideas in Alexandria Ocasio-Cortez's Green New Deal. *Business Insider*. Retrieved from: www.businessinsider.com/alexandria-ocasio-cortez-green-new-deal-support-among-americans-poll-2019-2?r=US&IR=T [Accessed 9 March 2019].
36 National Association of State Energy Officials. The 2019 U.S. *Energy & Employment Report*. Retrieved from: https://static1.squarespace.com/static/5a98cf80ec4eb7c5cd928c61/t/5c7f3708fa0d6036d7120d8f/1551849054549/USEER+2019+US+Energy+Employment+Report.pdf [Accessed 9 March 2019].
37 Wood Mackenzie, 2018. Thinking Global Energy Transitions: The What, If, How and When. Retrieved from: www.woodmac.com/reports/macroeconomics-risks-and-global-trends-thinking-global-energy-transitions-the-what-if-how-and-when-23699/ [Accessed 17 January 2019].
38 Pirani, S., 2018. *Burning Up: A History of Fossil Fuel Consumption*. London: Verso Books.
39 Realising Transition Pathways. EPSRC 'Realising Transition Pathways' Consortium: Key findings. Retrieved from: www.realisingtransitionpathways.org.uk/realisingtransitionpathways/publications/RTP_Achievements_One_Pager_Final.pdf [Accessed 20 August 2018].
40 Mazzucato, M., 2018. *The Value of Everything: Making and Taking in the Global Economy*. London: Allen Lane.
41 Mazzucato, M. 2018, *The Value of Everything: Making and Taking in the Global Economy*. London: Allen Lane.
42 Perez, C., 2016. Capitalism, Technology and a Green Global Golden Age: The Role of History in Helping to Shape the Future. *The Political Quarterly*, [online], 86 (1). Retrieved from: doi:10.1111/1467-923X.12240 [Accessed 17 December 2018].

43 Roberts, C. et al., 2018. The Politics of Accelerating Low-carbon Transitions: Towards a New Research Agenda. *Energy Research and Social Science*, [online], 44 (304–311). Retrieved from: doi:10.1016/j.erss.2018.06.001 [Accessed 17 December 2018].
44 Sovacool, B. K., 2016. How Long Will It Take? Conceptualizing the Temporal Dynamics of Energy Transitions. *Energy Research and Social Science*, [online], 13 (202–215). doi:10.1016/j.erss.2015.12.020 [Accessed 17 December 2018].
45 Laville, S., 2019, 16 January. Ministers to Review Durham Open-cast mine Decision. *The Guardian*. Retrieved from: www.theguardian.com/environment/2019/jan/16/ministers-to-review-durham-open-cast-mine-decision [Accessed 9 March 2019].
46 Turnheim, B. and Geels, F. W., 2012. The Destabilisation of Existing Regimes: Confronting a Multi-dimensional Framework with a Case Study of the British Coal Industry (1913–1967). *Research Policy*, [online], 42 (10). doi:10.1016/j.respol.2013.04.009 [Accessed 17 December 2018].
47 Inter-Governmental Panel on Climate Change (IPCC), 2018. Global Warming of 1.5°C. Retrieved from: https://report.ipcc.ch/sr15/pdf/sr15_spm_final.pdf [Accessed 20 November 2018].
48 Leipprand, A. and Flaschland, C., 2018. Regime Destabilization in Energy Transitions: The German Debate on the Future of Coal. *Energy Research and Social Science*, [online], 40 (190–204). doi:10.1016/j.erss.2018.02.004 [Accessed 17 December 2018].
49 Scoones, I. et al., 2015. The Politics of Green Transformations. *In*: Scoones, I. et al., eds. *The Politics of Green Transformations*. Abingdon: Routledge, 3.
50 Lockwood, M., 2013. The Political Sustainability of Climate Policy: The Case of the UK Climate Change Act. *Global Environmental Change*, [online], 23 (5). doi:10.1016/j.gloenvcha.2013.07.001 [Accessed 17 December 2018].
51 Roberts, C. et al., 2018. The Politics of Accelerating Low-carbon Transitions: Towards a New Research Agenda. *Energy Research and Social Science*, [online], 44 (304–311). Retrieved from: doi:10.1016/j.erss.2018.06.001 [Accessed 17 December 2018].
52 United States Climate Alliance. *Homepage*. Retrieved from: www.usclimatealliance.org/ [Accessed 9 March 2019].
53 Bernstein, S. and Hoffman, M., 2018. The Politics of Decarbonization and the Catalytic Impact of Subnational Climate Experiments. *Policy Sciences*, [online], 51 (2). https://link.springer.com/article/10.1007%2Fs11077-018-9314-8 [Accessed 17 December 2018].
54 Smith, A., 2012. Civil Society in Energy Transitions. *In*: Verbong, G. and Loorbach, D., eds. *Governing the Energy Transition: Reality, Illusion, or Necessity*. New York: Routledge, 157.
55 Köhler, J. et al., 2019. An Agenda for Sustainability Transitions Research: State of the Art and Future Directions. *Environmental Innovations and Societal Transitions*, [online]. Retrieved from: doi:10.1016/j.eist.2019.01.004 [Accessed 17 February 2019].
56 Huber, M., 2013. *Lifeblood: Oil, Freedom and the Forces of Capital*. Minneapolis: University of Minnesota Press.
57 Brand, U. and Wissen, M., 2018. *The Limits to Capitalist Nature: Theorizing and Overcoming the Imperial Mode of Living*. London: Rowman & Littlefield.

58 Klein, N., 2014. *This Changes Everything: Capitalism vs. the Climate*. New York: Allen Lane.
59 Looking Horse, C. A., 2018, 22 February. Standing Rock is everywhere: one year later. *The Guardian*. Retrieved from: www.theguardian.com/environment/climate-consensus-97-per-cent/2018/feb/22/standing-rock-is-everywhere-one-year-later [Accessed 20 November 2017].
60 Boyce, J. K., 2018. How Economic Inequality Harms the Environment. *Scientific American*. Retrieved from: www.scientificamerican.com/article/how-economic-inequality-harms-the-environment/ [Accessed 20 December 2018].
61 Roberts, C. et al., 2018. The Politics of Accelerating Low-carbon Transitions: Towards a New Research Agenda. *Energy Research and Social Science*, [online], 44 (304–311). Retrieved from: doi:10.1016/j.erss.2018.06.001 [Accessed 17 December 2018].
62 Downey, L., 2015. *Inequality, Democracy and the Environment*. New York: New York University Press.
63 Go Fossil Free. Commitments. Retrieved from: https://gofossilfree.org/divestment/commitments/ [Accessed 9 March 2019].
64 BBC, 15 February. Climate Strike: Schoolchildren Protest over Climate Change. *BBC*. Retrieved from: www.bbc.co.uk/news/uk-47250424 [Accessed 9 March 2019].
65 Chaplain, C., 25 February. Councils in England and Wales Declare 'Climate Emergency' Following UN Warnings of Future. *INews*. Retrieved from: https://inews.co.uk/news/environment/climate-emergency-declaration-councils-green-party-un-report/ [Accessed 9 March 2019].

Conclusions

This book has made the case that the richest people in the US and the UK have a unique role in contributing to climate change. It is time to recognise the reality of carbon inequality and gain a deeper understanding of how to tackle it. I have shown that the richest in the US and the UK have an *unequal ability to pollute.* The first component of this is luxury consumption. Their lifestyles are high-carbon because they are based on highly polluting forms of transport such as flying. Status competition with their peers means the richest keep up this lifestyle to ensure they are still recognised as a member of the elite. This has implications for policymakers if they ever moved to reduce the overconsumption of the richest. Asking rich people to voluntarily make their consumption more sustainable is unlikely to work because of intense status competition.

The second component of the elite's *unequal ability to pollute* is their *investment emissions* from holding shares in companies that produce significant greenhouse gas emissions such as oil, gas and coal. I identify members of the polluter elite in a database that accompanies this book. The database makes clear that *investment emissions* are much larger than consumption emissions. Therefore, the polluter elite are more responsible for global warming than citizens who only have consumption emissions. This raises moral issues because in addition to the richest holding greater responsibility, it is the poorest who suffer more, sometimes fatally, from air pollution and extreme weather events. Polluting activities such as oil and gas extraction which contribute to global warming and species extinction should no longer be defined as wealth creation.

The database also shows that *investment emissions* are not bound by territory. This has implications for comparing emissions between countries. It means that in addition to addressing the unequal distribution of emissions globally there needs to be specific recognition of the role of the *polluter elite* in the Anthropocene. It is not accurate to say that all of humanity has caused global warming. Instead, the historic role of the *polluter elite* in shaping the consumption options for the general public should be acknowledged. They have done this through their decisions to extract more oil, gas and coal and to seek political influence to defend the dominance of fossil fuels in the economy against rivals such as renewable energy.

The *polluter elite* who run the main oil and gas companies in the US and the UK understand that the state presents both a threat and an opportunity to their companies' ability to continue their polluting operations. Therefore, they lobby state policy as a matter of high importance. Ultimately, whether they are successful will affect their own personal net worth and elite status in society. To fully appreciate the polluter elite have been so politically effective it is important to look at how they interact with policymakers and the role of the state. This is not a one-way process as the revolving door between companies and government illustrates. It is vital to understand that the *polluter elite* have used successive governments' reliance on fossil fuels for economic growth as a lever to pressure them to uphold the fossil fuel status quo.

At key moments since 1990, the *polluter elite* have successfully blocked the acceleration of the low-carbon transition. The most recent and symbolic examples are President Trump's withdrawal from the Paris Agreement and the Conservative government's legislation in 2015 to maximise oil and gas extraction in the UK. In addition, the continued subsidies worth billions of pounds for fossil fuels highlight the dominance of the *polluter elite* at the precise time when the transition to a low-carbon economy needs to be accelerated. Their political power continues to grow as low tax rates further concentrate resources in their hands.

This is why the *polluter elite* need to be destabilised to open up space for clean alternatives. There has been a dramatic fall in costs for solar and wind generation which has seen an increase in installation of these technologies in the US and the UK. But they have not replaced the fossil fuel-based economy. If the *polluter elite* are not destabilised they will continue to use their wealth to lobby the state to maintain fossil fuels in the energy mix alongside renewable energy.

In 2019, the low-carbon transition appears within reach. Renewable energy is increasingly competitive with fossil fuels and there is greater public consciousness about climate change. However, if the wealth and political power of the oil and gas *polluter elite* are not challenged, then the state in countries such as the US and the UK will not reduce greenhouse gas emissions fast enough as part of global efforts to avoid a 1.5 degree increase in average surface temperature.

Policymakers could deliberately and immediately destabilise the *polluter elite* through targeted policies such as ending fossil fuel subsidies, introducing a carbon tax and phasing out fossil fuels. These measures would help accelerate the low-carbon transition but will only happen if there is overwhelming public pressure. There are signs this is growing with school children and the wider public in both the US and the UK starting to protest about climate change. They are frustrated with decades of inaction by their governments. This inaction and lack of political will is no coincidence. The *polluter elite* have used their economic and political influence to undermine legislation to reduce emissions. To accelerate the transition away from fossil fuels at the rapid speed that is necessary requires recognising the power of the *polluter elite* and directly challenging it.

Appendix 1

Ways in which the decision makers in oil and gas companies in the US and the UK seek political influence

Ways to obtain political influence	*Indicative examples from the United States*
Via their company they approve funding of political parties	For oil and gas companies these donations reach millions of dollars in each election cycle, almost always the Republicans.[1] On its website, ExxonMobil explains that the board reviews all political contributions and the company's lobbying activities on an annual basis.[2]
Via their company they approve funding of fossil fuel lobbyists	ExxonMobil disclosed that between January and December 2017 it spent $3.4 million on lobbyists to lobby the House and Senate.[3]
On behalf of their company they are members of fossil fuel industry lobby groups	Former CEO of ExxonMobil Rex Tillerson was an executive member of the American Petroleum Institute.
Via their company they approve climate change denial and sometimes make public statement questioning climate science	ExxonMobil's public position consistently cast doubt on climate change despite its own scientists' peer-reviewed research and internal memos acknowledging that global warming was caused by human activity in the 1970s.[4] It is alleged that between 2006 and 2015 the company and its associated foundation spent around $14.2 million on organisations that oppose climate science and policy measures to reduce emissions.[5] Thomas Fanning of the Southern Company said in an interview with CNBC in 2017 that he did not believe carbon dioxide emissions were a driver of climate change.[6]

(Continued)

Ways to obtain political influence	*Indicative examples from the United States*
	The Global Climate Coalition operated between 1989 and 2001 and brought together a number of mainly US-based oil, coal, chemical and car companies to question climate science. Its membership included several of the companies listed in the polluter elite database such as Chevron, Shell and ExxonMobil.
They make personal donations to political parties	US citizens overwhelmingly donated to the Republic party (see the polluter elite database for details).

Decision makers of polluting companies (owners, executive team and directors) in the UK also seek to influence the political process in similar ways

Ways to obtain political influence	*Indicative examples from the United Kingdom*
Via their company they approve funding for political parties (almost always the Conservatives)	Direct funding was more difficult to identify than in the US. The Conservatives received £15,500 in May 2005 and £7,500 in October 2005 from Kerr McGee Oil, UK. Kerr McGee North Sea donated £7,500 in 2004. Meanwhile, Labour received £65,000 in February 2017 from Balmoral Group Holdings (specialises in offshore services for oil and gas).[7]
	The fossil fuel companies have donated to All Party Parliamentary Groups (APPG), which operate as a space within Parliament to influence MPs and Peers. The APPG on Unconventional Oil and Gas which operated between 2013 and 2018 received funding from INEOS, Shell, Dow Chemical, Essar Oil and several "big six" energy companies via the Energy and Utilities Alliance.[8]
	Whilst not mentioned in the polluter elite database the main energy utility companies have a history of political donations, particularly when that political party was in government. Scottish Energy gave £48,400 to the Conservatives between 2011 and 2017, £128,562 to Labour between 2001 and 2016. The company gave £23,806 to the SNP between 2002 and 2003.[9]
	EDF energy gave £33,500 to the Conservatives between 2003 and 2016. Gave £12,000 to Labour between 2003 and 2005.[10]
	E.On gave £8,400 to the Conservatives in 2016 and gave £10,000 to Labour in 2007.[11]
	Drax Power Limited gave £9,600 to Labour in April 2018.

Ways to obtain political influence	*Indicative examples from the United Kingdom*
Via their company they approve funding of fossil fuel lobbyists	Several of the companies mentioned in this book employ lobbyists. The September–November 2018 lobbying register shows that Shell, Vitol, INEOS and Cuadrilla hired lobbyists.[12]
On behalf of their company they are members of fossil fuel industry lobby groups	Oil and Gas UK brings together the main companies working in this area. Many of them have interests in the North Sea. Several of the board members are in senior roles at companies listed in the polluter elite database including Chevron, ConocoPhillips, BP, Shell and Total.[13] Oil and Gas UK supports the APPG on British Offshore Oil and Gas Industry.[14] Another of these is UK Onshore Oil and Gas.[15] The Directors of UKOOG are from some of the main companies attempting to do fracking. They include Frances Egan, CEO of Cuadrilla Resources, Stephen Bowler, CEO of IGas, and Gary Haywood, Director of INEOS Upstream. UKOOG also formed the Natural Gas Coalition.[16]
They make personal donations to political parties	Compared to the US it is a lot less common for the executive team and directors at polluting companies to have personally donated to political parties (almost always the Conservatives). (See the polluter elite database for details).

Notes

1 Center for Responsive Politics. Oil and Gas: Top Recipients. Retrieved from: www.opensecrets.org/industries/recips.php?ind=E01 [Accessed 17 January 2019]. Center for Responsive Politics. Coal Mining: Top Recipients. Retrieved from: www.opensecrets.org/industries/recips.php?cycle=2018&ind=E1210 [Accessed 17 January 2019].

2 ExxonMobil. Political Contributions and Lobbying. Retrieved from: https://corporate.exxonmobil.com/en/Company/Policy/Political-contributions-and-lobbying [Accessed 17 January 2019].

3 ExxonMobil. Political Contributions and Lobbying. Retrieved from: https://corporate.exxonmobil.com/en/Company/Policy/Political-contributions-and-lobbying [Accessed 17 January 2019].

4 Supran, G. and Oreskes, N., 2017. Assessing ExxonMobil's Climate Change Communications (1977–2014). *Environmental Research Letters*, [online], 12 (8). Retrieved from: https://iopscience.iop.org/article/10.1088/1748-9326/aa815f [Accessed 17 February 2018].

5 ExxonSecrets. ExxonMobil Climate Denial Funding 1998–2014. Retrieved from: https://exxonsecrets.org/html/index.php [Accessed 17 January 2019].

6 Singer, M., 2018. *Climate Change and Social Inequality: The Health and Social Costs of Global Warming*. London: Routledge Books.

7 The Electoral Commission. Online Database Search. Retrieved from: http://search.electoralcommission.org.uk/ [Accessed 17 February 2019].
8 LilSis. All Party Parliamentary Group on Unconventional Oil and Gas. Retrieved from: https://littlesis.org/org/255514/All_Party_Parliamentary_Group_on_Unconventional_Oil_and_Gas [Accessed 17 February 2019].
9 The Electoral Commission. Online Database Search. Retrieved from: http://search.electoralcommission.org.uk/ [Accessed 17 February 2019].
10 Ibid.
11 Ibid.
12 Public Affairs Board. Public Affairs and Lobbying Register. Retrieved from: https://www.prca.org.uk/sites/default/files/PA%20register%20-%20Dec%20 2017%20Jan%202018%20Feb%202018.pdf [Accessed 17 February 2019].
13 Oil & Gas UK. Oil & Gas UK Board Members. Retrieved from: https://oilandgasuk.co.uk/wp-content/uploads/2019/02/Oil-Gas-UK-Board-Members-January-2019.pdf [Accessed 17 February 2019].
14 Oil & Gas UK. All Party Parliamentary Group. Retrieved from: https://oilandgasuk.co.uk/appg/ [Accessed 17 February 2019].
15 United Kingdom Onshore Oil & Gas UK, 2017. UKOOG Annual Report. Retrieved from: www.ukoog.org.uk/images/ukoog/pdfs/UKOOG%20Annual%20Review%202017%20-%20web3.pdf [Accessed 20 December 2018].
16 United Kingdom Onshore Oil & Gas UK. About the Coalition. Retrieved from: www.ukoog.org.uk/the-natural-gas-coalition [Accessed 17 February 2019].

Appendix 2

A brief history of key moments when the US and UK government have slowed down or accelerated the transition to a low-carbon economy. These decisions have affected the profitability of shares held by the polluter elite

United States		
Government	*Slowed the transition = profit for polluter elite*	*Accelerated transition = threaten profit of polluter elite*
George H. W. Bush (1989–1993) *Republican*		Signed up to the UNFCCC committing the country in principle to future emissions reductions.
Bill Clinton (1993–2001) *Democrat*		Signed up to the Kyoto Protocol.
George W. Bush (2001–2009) *Republican*	Withdrew from the Kyoto Protocol. Made fracking exempt from the Safe Drinking Water Act.	
Barack Obama (2009–2017) *Democrat*	Proposed cap-and-trade legislation defeated. Prior to 2008 and 2012 elections Obama promised to keep petrol prices low. Enabled offshore drilling in the Arctic in 2015.	Signed up to the Paris Agreement. Clean Power Plan introduced to cut emissions from electricity generation. Regulation on leasing for oil and gas operations on federal lands. Ruled against Keystone XL and Dakota access pipelines. Banned new oil and gas drilling in the Arctic and Atlantic in 2018.

(*Continued*)

United States

Government	*Slowed the transition = profit for polluter elite*	*Accelerated transition = threaten profit of polluter elite*
Donald Trump (2017-current) *Republican*	Withdrew from the Paris Agreement. Rolled back the Clean Power Plan as well as other EPA regulations.	

Source: Author.

United Kingdom		
Government	*Slowed the transition = profit for polluter elite*	*Accelerated transition = threaten profit of polluter elite*
John Major (1990–1997) *Conservative*		Signed up to the UNFCCC committing the country in principle to future emissions reductions.
Tony Blair (1997–2007) *Labour*	Abandoned fuel duty rise in 2001.	Signed up to the Kyoto Protocol.
Gordon Brown (2007–2010) *Labour*		Climate Change Act introduced legally binding target to reduce emissions by 80% by 2050 (carbon budgets monitored by the Climate Change Committee). Energy Act 2008 led to introduction of Feed-In Tariff in 2010.
David Cameron (2010–2016) *Conservative Coalition with Liberal Democrats (2010–2015)*	Infrastructure Act committed the government to maximising economic recovery of UK petroleum. Creation of Oil and Gas Authority to oversee this. The Energy Act enshrined the scope of the OGA as an 'arm's length' agency.	Signed up to the Paris Agreement. Carbon price floor taxed fossil fuels used to generate electricity. Announced plans to phase out coal-fired plants by 2025.

Government	*Slowed the transition = profit for polluter elite*	*Accelerated transition = threaten profit of polluter elite*
Theresa May (2016-current) *Conservative*	Relaxation of planning laws to enable fracking. Overrule local councils to enable fracking to go ahead. In 2018 announced end of subsidy for renewable energy via Feed-In Tariff.	Launched Powering Past Coal Alliance with Canada. Requested the Climate Change Committee to advise on how the UK could achieve net zero emissions.

Index

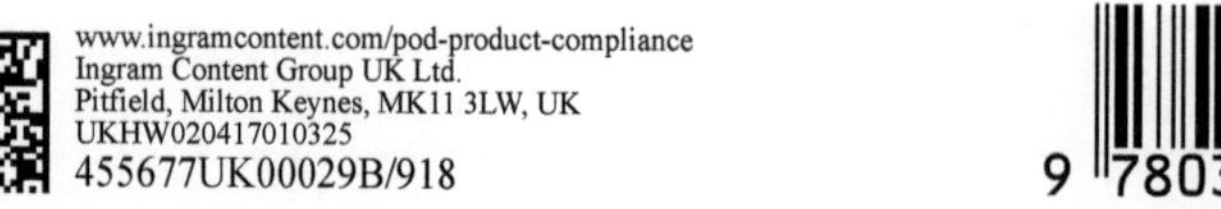

www.ingramcontent.com/pod-product-compliance
Ingram Content Group UK Ltd.
Pitfield, Milton Keynes, MK11 3LW, UK
UKHW020417010325
455677UK00029B/918

9 780367 727666